FLY FISHING GUIDE TO VERMONT

FLY FISHING GUIDE TO VERMONT

Complete Guide to Locations, Hatches, and History

MIKE VALLA

STACKPOLE BOOKS

Essex, Connecticut
Blue Ridge Summit, Pennsylvania

STACKPOLE BOOKS
An imprint of The Globe Pequot Publishing Group, Inc.
64 South Main Street
Essex, CT 06426
www.GlobePequot.com

Photos by Mike Valla unless otherwise noted

British Library Cataloguing in Publication Information available

Library of Congress Cataloging-in-Publication Data available

ISBN 978-0-8117-7691-2 (paper : alk. paper)
ISBN 978-0-8117-7776-6 (electronic)

Printed in India

Contents

Acknowledgments . . . . *ix*
Introduction . . . . *xi*
State of Vermont Hatches . . . . *xxi*

HUDSON RIVER BASIN **2**
Batten Kill 2
Green River 16
Roaring Branch (of Kelly Stand/Arlington) 17
Branch Pond, Bourn Pond, and Grout Pond 20
Walloomsac River 22
Roaring Branch (of Walloomsac), City Stream, and Bolles Brook 26
Hoosic River 29

LAKE CHAMPLAIN BASIN **36**
Lake Champlain—Middle Section 36
Mettawee River 40
Flower Brook 47
Poultney River 49
Castleton River 52
Lake Bomoseen 55
Lake St. Catherine 58
Big Branch Stream 59
Otter Creek 62
East Creek 66
Furnace Brook 67
Neshobe River 69
Sugar Hill Reservoir (Goshen Dam) 72
Cold River 73
Mill River 74
Middlebury River 76
Lewis Creek 78
New Haven River 82
Winooski River 85
Huntington River 93
Little River below Waterbury Reservoir 95
Little River—West Branch above Waterbury Reservoir 97
Ranch Brook 100

Mad River 101
Dog River 105
Lamoille River 109
Lamoille River—North Branch 117
Gihon River 119
Browns River 121
Missisquoi River 123
Missisquoi River—East Branch 124
Missisquoi River—Main Stem 126
Lake Champlain North Section at Missisquoi Bay Area 131
Trout River 134
Trout River Tributaries—Wade Brook and Jay Brook 136
Trout River—South Branch 138

MEMPHREMAGOG BASIN 142
Pherrins River 143
Jobs Pond 148
Bald Hill Pond 149
Long Pond 150
Lake Willoughby 151
Willoughby River 154
Shadow Lake 156
Echo Lake 157
Clyde River 160

CONNECTICUT RIVER BASIN 170
Forest Lake/Averill Lakes—"Quimby Country" 170
Upper Connecticut River 175
Nulhegan River 184
Notch Pond 188
Paul Stream 190
Moose River 192
Passumpsic River 195
Passumpsic River—East Branch 196
Passumpsic River—West Branch 197
Passumpsic River—Main Stem 199
Joe's Brook 201
White River 202
Black River 209
Jewell Brook 212
Bailey Brook 214

Knapp Pond #1 and Knapp Pond #2 215
Wells River 217
Noyes Pond (aka Seyon Pond) 220
Martins Pond 223
Waits River 225
Ompompanoosuc River 226
Ompompanoosuc River—East Branch 226
Ompompanoosuc River—West Branch 228
Ottauquechee River 230
Colton Pond and Kent Pond 232
Deerfield River 234
West River 245
Jenny Coolidge Brook and Greendale Brook 246
Williams River 248
Saxtons River 252

SELECTED FLY PATTERN RECIPES 256

Index 263
About the Author 272

Acknowledgments

As I've said many times over the years, books of this type can't be completed without help from many people—not just fly anglers. Fisheries biologists from Vermont Fish & Wildlife (VF&WD) provided a tremendous amount of information pertaining to the state's waters. I can't say enough good things about Vermont's fisheries professionals. Thanks go to Jud Kratzer, Lee Simard, Bret Ladago, Shawn Good, Courtney Buckley, and Lael Will. I've turned to Green Mountain National Forest biologist Scott Wixsom more than a few times over the years for information pertaining to his service area. Thanks also go to Joe Craine and the scientists at Jonah Ventures DNA Laboratory. Others include Clark Amadon, Steven Atocha, Ted Benoit, Morgan Boyd, Paul Bugeja, Brian Comfort, Willy Dietrich, Greg Fay, Shane Hanlon at the Dwight D. Eisenhower Federal Fish Hatchery, Jeff Haylon, Ralph Kucharek, Tim Major, Adriano Manocchia, Colin Phillips, Drew Price, Nick Proulx, John Rogers, Stan Steen, Scott Sztorc, Leighton Wass, Dean Wheeler, and Parker Wright. Flies tied by Bob Adams, Tom Baltz, Scott Biron, Pat Cohen, Ed Collins, John Collins, Bob Herson, Dave McNeese, Tom Miller, Al Pitt, Henry Ramsay, George Schlotter, and Rich Strolis appear in these pages.

Special thanks to my wife, Valerie, who once again made a book possible. There was never a question when I decided to head out the door to Vermont in search of waters and fish. Sometimes it was Val's prodding that got me motivated to keep things moving along. Wading through streams, hiking long trails, and paddling around ponds, she was always a willing participant. The eyes behind a camera lens, Val shot many of the scenic and action photos.

Many thanks to the entire Stackpole team: editors Jay Nichols, Meredith Dias; copy-editor Paulette Baker; proofreader Kris Patenaude; cover designer Amanda Wilson; and layout artist Karen Weldon.

Vermont is synonymous with blazing fall foliage. Brilliant yellows, reds, oranges, and golds create backdrops for both leaf peepers and anglers. VALERIE VALLA

Introduction

When asked about their impressions of Vermont, most would mention almost anything but its fishing opportunities. The state is most synonymous with its fall foliage that attracts leaf peepers, who travel from afar to view the yellows and reds and oranges that blaze the countryside. Abundant sugar maples that help create that attraction are also responsible for the state's renowned maple syrup industry. Sugarhouses, as they're called, are found everywhere in Vermont, even along the most remote dirt and gravel country roads. Their cupolas billow steam from maple sap being boiled in evaporators, an activity that happens anywhere from March into April.

Vermont's many quintessential and even historic covered bridges are something else you might have heard about, for good reason, since more than 100 cross the streams all over Vermont. You'll even hear about Vermont's numerous first-class ski resorts like Killington and Stowe Mountain Resort that attract winter sports enthusiasts; skiing is a major component of Vermont's economy, which is largely focused on recreational tourism. Trout streams flow in close proximity to the resorts. Mountain biking along Vermont's trailways has taken off in recent years too. Yet hidden among all the other enjoyable recreational activities, tourism, and features, Vermont is also known for its enticing fly-fishing opportunities. Fly fishing is popular on Vermont's waters, and Vermont has active Trout Unlimited chapters throughout the state. The Native Fish Coalition is also active in Vermont, as are various other conservation groups that focus attention on the state's fisheries and waters.

When comparing fly-fishing opportunities available in Vermont, the 45th largest state, with adjoining New York State (the 27th largest state), one might falsely assume fly-fishing prospects are minimal. While it's true most of Vermont's trout rivers are not as fertile as streams that flow through New York, Pennsylvania, or elsewhere, for a combination of reasons we'll discuss in the Hatches section that follows, there's still plenty of interesting fishing to experience in the Green Mountain State.

Vermont is blessed with many ponds, lakes, and reservoirs—nearly 200 are listed in Vermont's official State Fishing Regulations document. The document also lists 64 notable rivers and streams. Those numbers don't include the vast number of trickles that flow through the mountains. All told, Vermont provides fishing for rainbow, brook, and brown trout and many other fish species that can be taken on flies. You won't find the vast number of steelhead waters found in other states such as New York, yet steelhead fishing is available on several notable Vermont streams, including Lake Champlain tributaries and those in the eastern part of the state. Lake-run landlocked salmon provide some of the most interesting fishing in the state, and the history that surrounds landlocked salmon fisheries in Vermont is fascinating.

Warmwater species flourish in Vermont, including everything from bass to pike to the rare redfin pickerel. VF&WD manages and studies many of these waterbodies and ensures public access to ponds and lakes via many formal boat launches. If there's a particular pond or lake you might be interested in exploring, chances are a public launch is available, always marked by clear signage. Have you ever caught a "prehistoric fish" on a fly? If not, Vermont's share of Lake Champlain should be on your list. Bowfin, a surviving species of the Early Triassic epoch, 250 million years ago, inhabit the lake's shallows. Lake trout can be enticed with flies on Lake Champlain especially during early spring. Gar as large as an arm, and longer, also swim in Champlain's waters.

The author caught this bowfin on a shallow backwater area near mouths of tributaries that run into Lake Champlain while fishing with notable bowfin fishing expert Captain Drew Price. The bowfin gobbled a Mr. Bow Jiggles fly dangled down through buttonbush cover to the fish's snout. DREW PRICE

Both largemouth and smallmouth bass inhabit many of Vermont's ponds, lakes, and rivers. Waters such as Lake Bomoseen contain not only average-size fish; multi-pound largemouth are taken every year.

Vermont stocks substantial numbers of rainbow trout throughout Vermont. Rivers such as the White are known for the spunky rainbows that swim through its fast-moving waters.

Even more fascinating than what can be experienced in the lowlands, in my opinion, is the number of brook trout fishing opportunities available along high-elevation small creeks and brooks that run through the Green Mountain National Forest. As one Vermont fisheries biologist recently told me, the vast number of these little brooks are underutilized by anglers. Vermont's wild brook trout that inhabit the forest flows are not large (with exceptions). Generally, a dollar bill–size brookie is a real gem; their beauty alone is enough to make up for length.

Wild brook trout are gems that inhabit many of Vermont's headwater trickles. Organizations such as the Native Fish Coalition, VF&WD, USDA Forest Service, and Trout Unlimited work together to help ensure their presence in Vermont's waters.

Most any colorful wet fly will take brook trout inhabiting high-elevation streams. Small dry flies also work well.

Small mountain stream tributaries like Ranch Brook provide fishing for small yet colorful wild brook trout in a woodland setting.

Wild brook trout that range in size from young-of-the-year to dollar-bill size inhabit numerous high-elevation brooks throughout Vermont. True jewels of Vermont, the fish are often small but provide fun sport.

Colorful brown trout inhabit many waters throughout Vermont. Rivers such as the Batten Kill are known for browns, but they also swim in many of the smaller creeks throughout Vermont.

My own experiences fly-fishing Vermont began many years ago, in the late 1970s. My wife Valerie's family resides in eastern New York State near Schuylerville, a short drive from Vermont. Even before we moved to Saratoga Springs, New York, in the late 1980s, we were making frequent jaunts to visit Val's parents and family. Of course those trips called for a bit of escape into Vermont to fish streams such as the Castleton, Mettawee, and Walloomsac Rivers and the Batten Kill. It didn't take long for me to figure out what was also available along brook trout streams in the Green Mountains.

For a while we owned a small fishing cottage in Shushan, New York, within walking distance of the Batten Kill and just a couple of miles downstream from the Vermont border. Once we moved to the town of Cambridge, New York, also a quick drive to the Vermont border, my explorations continued. We now live in a log cabin on a quiet countryside road nested in mostly dairy country. Our morning walks take us up our road to a vantage point, where we often briefly stop to view the ridgeline of Vermont's Green Mountains off in the distance, north of Bennington. Some of my favorite forest brook trout streams are in that area.

Vermont's northeast corner areas, and what fly-fishing opportunities existed up that way, remained largely a mystery. I had heard stories from angling friends about the awesome Northeast Kingdom and the spectacular lakes, ponds, and streams up that way. What is the Northeast Kingdom? Former governor George D. Aiken is credited with first using the term "Northeast Kingdom" (NEK to many) during a 1949 speech. In May 2014, then-governor Peter Shumlin signed a law designating the Governor Aiken Bucktail the official fishing fly of the state of Vermont. The bucktail, often tied and fished in many tandem and variation versions today, has been around for decades. Joseph D. Bates Jr. helped popularize the pattern in his now classic *Streamer Fly Tying and Fishing* (1950). The streamer was popular in the Lake Memphremagog region for landlocked salmon and rainbow trout; it's still is use today.

Governor Aiken Streamers *Tied by Scott Biron*

Well-known New Hampshire streamer fly tier Scott Biron crafted these popular versions of Vermont's official state fishing fly. The streamer is still in popular use on many ponds and lakes in Vermont's Northeast Kingdom.

Today you'll see that label—Northeast Kingdom—attached to everything from eateries to car washes when driving through areas such as St. Johnsbury and Newport on your way to fish those surrounding waters. Bounded roughly by the Green Mountains to the west and the Connecticut River to the east—in Orleans, Caledonia, and Essex Counties—the "kingdom" is accentuated by large oligotrophic lakes like Willoughby and Echo, as well as rivers such as the Clyde and Nulhegan. Notable streams contain trout in the Northeast Kingdom—Nulhegan River, Paul Stream, Pherrins River, and the Clyde—making a long drive to that part of Vermont not far from the Canadian border worth the time and effort.

DNA Analysis of Vermont's Streams

While there's not much mystery what fish species swim through waters such as Lake Willoughby, Echo Lake, and others, sometimes it's not all that simple to ascertain what species inhabit some of the smaller creeks and brooks. Although I've fished many Vermont waters over the years, I don't always net the full range of fish species that might inhabit a given water. You'll notice throughout these pages frequent mention of taking eDNA samples on many streams throughout Vermont, particularly along headwater tributaries, where surprises might exist.

I'm fortunate to own a Smith-Root eDNA sampling unit. Smith-Root has been known for its electrofishing devices, widely used by fisheries biologists for many years. Since college genetics courses, I've always had an interest in this facet of biology. Joe Craine and his group of scientists at Jonah Ventures DNA labs in Colorado provided superb analysis of the numerous eDNA samples I submitted.

The author obtaining a DNA sample at the headwaters of a Vermont stream that rainbow trout were alleged to inhabit but not absolutely proven. One hundred percent base-pair match came through for *Oncorhynchus mykiss*, rainbow trout. VALERIE VALLA

Through DNA sequencing, a given sample analysis will show virtually every fish species inhabiting a given water. In a stream analysis data sheet result, I looked for 100 percent base-pair matches for salmonids in particular, although many other fish species and other vertebrates show up in a sample. One of Vermont's fisheries biologists was aware I was doing lots of DNA sampling around the state. He had heard anecdotal talk of rainbow trout possibly appearing in an upper tributary where that species had never been detected with electrofishing surveys. He asked me to take a sample at a specific GPS coordinate if I was up that way on the stream. When the analysis of the sample DNA arrived, there wasn't a lot of it, but 100 percent base-pair markers for *Oncorhynchus mykiss*, rainbow trout, came through.

Because of the enormous undertaking required to provide as much detail as possible about some streams I was less familiar with, I hauled both fly rods and my Smith-Root DNA sampler on many Vermont outings. DNA findings on some waters were sometimes surprising. I certainly never expected a sample to turn up *Sus scrofa*—wild boar. Vermont wildlife biologists were actively interested in where this invasive animal had wandered around the state and solicited sightings, so I forwarded my findings to the VF&WD.

Fishing Regulations Updates

Beginning January 1, 2022, Vermont changed the open-season fishing dates. The new regulations allow fishing in almost all waterbodies throughout the state year-round, so long as anglers practice catch-and-release. However, always check the official Vermont Fishing Regulations, available online and in printed form. Prior to the changes, fly fishers who were anxious to get into their waders and cast for early-season trout had to wait until mid-April. Anglers now can experience very early season fishing in March, about the time the sugarhouses start boiling maple sap to make syrup.

Fishing regulations now allow fly fishers to get into their waders and cast streamers into waters year-round. Cold-hardy anglers fishing March into April can observe steam billowing up from the hundreds of maple syrup sugarhouses throughout Vermont.

Vermont Trophy Trout Waters

Vermont has designated some 11 rivers and 36 lakes and ponds around the state as Trophy Trout Waters. Larger size trout are stocked annually in these waters. While some of the fish hold over, many are primarily "put-and-take" waters, where heavy harvesting during early season by anglers is expected. Among these streams are Black River in the Cavendish area, Deerfield River in Searsburg, East Creek in Rutland City, Lamoille River in Fairfax, Little River in Waterbury, Otter Creek near Danby, Missisquoi River in Ensoburg, Passumpsic River in St. Johnsbury, Moose River in St. Johnsbury, Walloomsac River in Shaftsbury and Bennington, and Winooski River in Waterbury and Duxbury.

General Precautions

Use common sense while fishing the many swift rivers around Vermont—not just at hydroelectric facility reservoir discharges and tailraces. Be aware that water levels can unexpectedly rise to dangerous levels. Don't wade too deep, and always have an escape plan that will quickly get you out of the water and out of danger. An angler drowned a couple of years ago at Peterson Dam on lower Lamoille River. Rivers such as the Huntington and its gorge have taken many lives over the years. Observing another angler fishing a river stretch doesn't mean it is safe for others to do so.

Vermont has a number of hydroelectric facilities that provide fishing opportunities below dams. Anglers should always heed the signage placed along the waters.

Other precautions worth noting include trips into isolated areas in the Green Mountain National Forest. If available, always sign in at trailhead kiosk stations. Make sure others know your plans and when you're expecting to return home. A good practice if enjoying such an experience with others is to plan your extended trip around the weakest member of the party.

Useful Tools

The following resources can prove useful in your fly-fishing endeavors:

- **Topological maps:** High-scale USGS topological maps are very useful tools that are now easily accessible online. Topo maps are particularly useful for locating many small high-elevation tributaries throughout Vermont that contain wild brook trout: https://apps.nationalmap.gov/downloader/.
- **Lower scale Vermont topo maps** are available in *DeLorme: Vermont Atlas & Gazetteer*, a great companion while exploring areas around the state. Compiled in book form, the maps indicate everything from boat launch locations to trailheads.
- **VF&WD Annual Fisheries Assessments:** Vermont's Fish & Wildlife Department has an interesting tool that enables anglers and others to view a large number of their annual fisheries reports. After accessing the site, simply indicate the waterbody you are interested in learning more about. Fish species population assessments are usually

included in the reports, along with other pertinent information: https://anrweb.vt.gov/FWD/FWLibraryExternal/SearchFisheriesLibrary.aspx.

- **Fishing Access Areas:** VF&WD maintains an online site with ample information about various fishing access sites throughout the state. Nearly 200 formal access areas can be located by indicating waterbody, county, boat size, and fish species. Statewide maps are also available at the site: https://vtfishandwildlife.com/fish/boating-in-vermont/fishing-access-areas.

Rainbow Trout Strain Evaluation

The final year of VF&WD's three-year rainbow trout evaluation study was 2024, and results were not available at the time of publication. Twelve waterbodies were selected to compare two strains—one a new hatchery strain developed from a species that had ancestors in deep and cold Eagle Lake. That strain was compared with the Erwin-Arlee strain. The strain more frequently caught by anglers will be the one stocked along Vermont's waterbodies. You might still encounter informational signage posted along the streams that were stocked for the comparative study.

Connecticut River Regulations

The Connecticut River borders New Hampshire, so make sure you read and understand any special regulations that apply, especially to nonresidents of Vermont and New Hampshire. Review both the official Vermont Fishing Regulations and New Hampshire Fishing Regulations, available online and in print form, for particulars concerning fishing the Connecticut River. You'll want to obtain both nonresident New Hampshire and Vermont licenses if you're not a Vermont resident. Some of the best upper Connecticut River water flows through New Hampshire, where different regulations apply.

Enjoy the Full Experience

A point I made in my *Fly Fishing Guide to New York State* (2023) deserves mention here also. Like New York—or any other state for that matter—there's a heck of a lot more to experience after climbing out of your waders. Take time to read the many state and local historic markers scattered along Vermont's streams and trailheads. You'll be inspired if not amused by the likes of the marker placed next to the Hoosic River in North Pownal that tells the story of an alleged witch who was thrown into the river. And of course there's the famous Emily's Bridge, a covered bridge over Gold Brook where her ghost is said to appear. The covered bridges alone are enough to grab your attention when you're not reeling in a trophy fish, with perhaps a fall mountain scene, worthy of an artist, as backdrop.

Silvio O. Conte National Fish and Wildlife Refuge, along the Nulhegan River in the Northeast Kingdom, is major attraction during the colorful fall foliage display of late September into October.

As you trek along many of the streams, you'll come across such interesting sites as remnants of old mills like the one that once existed on the Ompompanoosuc River, the site of an 1800s ironworks along the Middlebury River, or even remains of entire communities that have vanished, such as the one at the headwaters of Joe's Brook near St. Johnsbury. The Silvio O. Conte National Wildlife Refuge visitor center is a must-see while fishing the Nulhegan River in the Northeast Kingdom. Suffice it to say that there's much more than fish to enjoy in and around Vermont.

As you travel through the state, please consider patronizing the various small eateries, shops, and stores that have endured devastating effects of several high rain and flood events that occurred in 2023 and twice again in 2024. Some streams have still not recovered from 2011's Tropical Storm Irene. You'll no doubt view just how catastrophic those events were; many river valleys, including roads and infrastructure, were wiped out. Some businesses survived the 2023 event only to be knocked down again in 2024.

Vermonters are friendly sorts. Valerie and I patronized many of the tiny diners tucked away in small towns throughout the state, especially in the Northeast Kingdom. The folks who shared a meal with us were always helpful when asked about fishing in their areas. They'll share nice stories about their towns, not just about fishing, even how to properly pronounce river names of native origin that are often more than a mouthful—rivers such as the Ompompanoosuc.

A historical marker at the Alan R. Strobridge Recreational Complex along the Hoosic River in North Pownal tells the tale of the famous witch trial ca. 1785.

State of Vermont Hatches

Many of the same mayflies you'll encounter on New York's Beaver Kill or Pennsylvania's Penns Creek emerge from Vermont streams, but not in the same prolific numbers. It all has to do with bedrock geology, along with other abiotic factors such as acidic conditions, stream geomorphology, and streambed characteristics. Taxa diversity and density of macroinvertebrates around cobble and small stones is much greater than found in systems with sand substrate. I recall when my faculty advisor at Cornell University, Dwight A. Webster, was scratching out a list of courses he insisted I enroll in. Geology? *Why geology?* I wondered. I'm there to study fisheries biology. A couple of Vermont locals I communicated with pointed to human-induced changes that no doubt also have something to do with macroinvertebrate declines experienced in more recent times; the decrease in riparian vegetation and concomitant sedimentation issues are probably in the mix of reasons, along with the recent high rain and flooding events. Of course there are exceptions here and there across the state. You'll encounter storms of Grannom (*Brachycentrus* spp.) hatches on Batten Kill, and even way up north on the Nulhegan River in Vermont's Northeast Kingdom. Batten Kill, my home water, has great Hendrickson (*Ephemerella subvaria*) hatches—probably the most anticipated hatch on that stream. Every August I wait in anxious anticipation for *Tricorythodes* to appear on Arlington-area flats. There was an occasion a few years ago when I encountered a significant Sulphur (*Ephemerella invaria*) spinner fall near the West Branch of Little River—the backwater was covered with them.

Hexagenia limbata is no doubt the most anticipated big mayfly emergence throughout Vermont. Small ponds half-hidden in the Green Mountain National Forest to larger oligotrophic ponds and lakes in the Northeast Kingdom have plenty to get anglers stirred up about from late June into July.

GENERAL VERMONT STATE HATCHES EMERGENCE DATES

Common Name	Latin Name	General Emergence Ranges Early to Late	Sizes
Little Black Stonefly	*Taeniopteryx nivalis*	Late February to early April	#18-20
Early Brown Stonefly	*Strophopteryx fasciata*	Early March to early April	#14-16
Little Blue-Winged Olive	*Baetis*	1st: March to April 2nd: Late September to October	#18-20
Little Black Caddis	*Chimarra* spp.	Mid-April to early September	#16–18
Spring Blue Quill	*Paraleptophlebia adoptiva*	Mid-April to mid-May	#16-18
Grannom	*Brachycentrus* spp.	Late April to early May	#12-14
Apple Caddis	*Brachycentrus appalachia; B. spinae*	Early May to late May	#14-16
Hendrickson/Red Quill	*Ephemerella subvaria*	Mid-April to mid-May	#14-16
March Brown	*Maccaffertium vicarium*	Mid-May to mid-June	#10-14
Big Sulphur	*Ephemerella invaria*	Mid-May to late June	#14-16
Giant Stonefly	*Pteronarcys dorsata*	Late May to late June	#6-8
Golden Stonefly	*Isoperla* and *Acroneuria* spp.; *Paragnetina immarginata*	Late April to late July	#6-12
Dark Blue Sedge	*Psilotreta labida; P. frontalis*	Mid-May to late June	#12-14
Little Sulphur	*Ephemerella dorothea dorothea*	Late May to late June	#18-20
Brown Drake	*Ephemera simulans*	Late May to late June	#10-12
Blue-Winged Olive	*Drunella lata*	Early June to September	#14-16
Cream Cahill	*Maccaffertium* spp.	Early June to late June	#14
Dun Variant (Slate Drake)	*Isonychia bicolor*	1st: Late May to late June 2nd: Mid-September to early October	#10-14
Yellow Sally, Lime Sally	*Isoperla; Alloperla*	Mid-May to mid-July	#14-16
Light Cahill	*Stenonema interpunctatum*	Mid-June to mid-July	#14
Dark Green Drake	*Litobrancha recurvata*	Late May to late June	#8-10
Hex Drake	*Hexagenia limbata*	Late June to mid-July	#8-10

Yellow Drake	*Ephemera varia*	Late June to mid-July	#10-12
Golden Drake	*Anthopotamus distinctus*	Late June to mid-July	#10-12
Trico	*Tricorythodes* spp.	Mid-July to mid-September	#22-26
Hebe, Pale Evening Dun	*Leucrocuta hebe*	Early August to early October	#16-18
October Caddis	*Pycnopsyche* spp.	Early October to late October	#12

Spinner falls like *Ephemerella invaria* can be substantial on some Vermont streams, including smaller waters. These were encountered washed up near the West Branch of Little River, near Ranch Brook.

Male Hendrickson (*Ephemerella subvaria*) hatches are enjoyed on Vermont's Batten Kill and other streams during April into May.

Good Yellow Stonefly (*Isoperla* and *Alloperla*) flights can be observed on some streams, including the headwaters of Missisquoi River during May.

Various Blue Wing Olive mayfly species (including *Baetis* examples) are abundant enough on some of Vermont's streams to bring fish to the surface well into the fall months.

Roaring Branch along Kelly Stand Road tumbles down from the Green Mountain National Forest. Sporting artist Adriano Manocchia samples a small brook trout–inhabited pool nestled among boulders and rocks.

Hudson River Basin

Tucked down in Vermont's southwest corner, the Walloomsac River, Batten Kill, a short section of the Hoosic River, and their tributaries compose the smallest yet not insignificant waters of the Hudson River drainage basin. Although Vermont's Hudson River drainage basin has fewer fly-fishing ponds when compared with other drainage basins, a few offer some interesting fly-rodding experiences. Bourn Pond, an isolated Lye Brook Wilderness brook trout pond, provides backcountry fishing. Branch Pond, also a brook trout pond, is easily accessed and in the neighborhood of Bourn and Grout Pond. Bourn and Branch are surveyed and monitored regularly for acid-condition status. (Anglers who fish the upper Deerfield River headwaters of the Connecticut River Basin are also in close proximity to those ponds.) Lake Shaftsbury, a warmwater fishery, is located south of Arlington at Lake Shaftsbury State Park.

There's no doubt that the singular fishery in the basin of paramount importance to fly fishers is Batten Kill. While it's not without its myriad peripheral multiuse problems, it's a good trout fishery. The 2021 *Basin-1 Tactical Basin Plan* succinctly described reasons behind the robust fishery: good habitat. "This exceptional habitat is provided by comparatively high summer flows, limestone in the drainage, extensive gravel beds for spawning, good shade along the banks, stable banks only occasionally subject to serious erosion and sedimentation, consistently cold summer water temperatures, a well-vegetated watershed with minimal nutrient and sediment pollution, and a naturally reproducing Brook Trout and Brown Trout fishery. The river has stable flows, cool water temperatures favorable to salmonids, good physical habitat providing necessary cover, refuges, spawning and rearing habitats, as well as the food quality and quantity necessary to support a healthy salmonid population."

Of course Batten Kill itself isn't the only Hudson River Drainage Basin stream fishery that receives attention from fly fishers. Plenty of small Batten Kill tributaries harbor wild brook trout and brown trout. Just around the corner from Batten Kill are the Walloomsac and Hoosic Rivers.

Batten Kill

There's little argument that Batten Kill is one of the better-known trout streams in the Hudson River Drainage Basin. In many ways, the Batten Kill I first experienced in the late 1970s is a much different river than anglers might find today. During the late 1970s I fished the river often during frequent visits to my wife Valerie's family, who live a half hour from the New York reaches. By the time we moved to Saratoga Springs, New York, in 1988, I spent more time on New York's mileage and explored additional stretches that flow through Vermont. Even more time was devoted to fishing the entire Batten Kill, from its source to its end-waters, once we purchased a fishing cabin in Shushan that was within walking distance to the river.

Over the years I've witnessed more than few changes along the river. Practices that have enhanced the fishery, such as habitat improvements, were sorely needed. Habitat

Local sporting artist Adriano Manocchia fishing the Keelan reach in Arlington. Easy access to the lovely stretch is made possible by Bill Keelan, a community leader who once ran a bed-and-breakfast on the property.

improvement initiatives, something both Vermont and New York have been involved with along with organizations such as Trout Unlimited and other conservation groups, have proved the key reason for the Batten Kill's relatively recent upswing in maintaining a healthy wild trout population. My good friend and premier fly-fishing guide Scott Sztorc has netted some mighty fine large wild brown trout in the past few years. Scott lives in Shushan and devotes hundreds of hours a year fishing the river's myriad runs and pools. My fishing buddy and local sporting artist Adriano Manocchia and I have enjoyed similar success on our favorite beats.

However, the multiuse concept of a natural resource has led to a degree of frustration among many fly fishers. Commercial activities, such as the growing interest in recreational tubing during summer, have detracted from an otherwise pleasant angling day and experience on the river. However, most of that issue occurs along New York's mileage from the Vermont–New York border downstream into Shushan. Yet plenty of recreational kayaking and tubing takes place on Vermont's stretches near Arlington during summer.

Commercial enterprises shuttle recreational floaters to areas that were once good parking and access points for fly fishers. During some periods of the season, the river takes on a kind of theme park atmosphere. Batten Kill regulars, including this author, dodge some of this activity by limiting our times on the water to either very early morning or midevening. We have no issues with a friendly canoer or kayaker floating past us in a respectful way during the evening, offering a friendly wave as they pass.

Indeed, attention to the river and its usage, not only for fly fishing, has grown over the years—on all fronts. On the one hand, my feeling is the river doesn't need any more attention or added angling pressures while it is still healing. The river system is on a positive track, and fishing is better than it has been in years, due largely to the tremendous

During summer, some stretches of Batten Kill can be likened to a theme park. The multiple-use concept of a natural resource attracts other users, such as recreational floaters and swimmers. Batten Kill fly-fishing regulars have learned to avoid some river sections at certain times of the day during summer.

efforts devoted to habitat enhancement. On the other hand, to not include the Batten Kill in a Vermont fly-fishing guide makes little sense. With some reluctance, I'll recommend a few well-known "starter" access areas to consider when heading to fish the Batten Kill. There's much more that could be disclosed, but its best to leave the rest to the angler's own exploration and learning process.

For many of us, part of the thrill of fishing the Batten Kill is the challenges the river presents. In his 1993 book *The Battenkill*, John Merwin wrote that the river is "among the—if not the most—technically difficult fly-fishing streams in America." To land a big, wild Batten Kill brown trout on a size 16 dry fly is a big accomplishment by any standard. Despite other competing uses of the river, there's plenty of water to explore along its length, from its source to its confluence with the Hudson River.

The 59-mile Batten Kill, 28 miles of it in Vermont, begins its journey in East Dorset as a small brook trout stream forged from Mad Tom and Little Mad Tom Brooks. Near Manchester Center it picks up additional flows from its unassuming brook trout–filled West Branch, which arises from the northern slope of Bear Mountain in the town of Rupert. After picking up the West Branch, the main stem Batten Kill continues southwesterly to Arlington, then west another 10 miles or so to the New York border, accepting inflow from the Green River, a small tributary, along the way.

In New York the stream pushes into the little hamlet of Shushan, picking up Camden Creek before turning north and passing near Salem, where it gathers Black Creek and White Creek, both trout streams. The Batten Kill widens and maintains its low-gradient flow through East Greenwich, where it again changes direction, heading south toward Battenville and then west to Greenwich and Middle Falls, before it cascades over Dionondahowa Falls then surrenders to the Hudson River near Schuylerville, New York. Trout inhabit

VF&WD biologists annually survey trout populations using electrofishing techniques on the river in the Arlington area.

nearly all of New York's share of the Batten Kill as it winds and turns after clearing the Vermont border, a fact that may surprise nonregulars, who may be more familiar with the famous Vermont section. No matter what reach in which state you're planning to tackle, Manchester, Vermont, and neighboring Arlington should be in your plans.

MANCHESTER AREA

As far as Vermont's share of Batten Kill, the village of Manchester and town of Arlington are headquarters. Fly anglers anticipating a Vermont Batten Kill fly-fishing experience should head for Manchester or the town of Arlington first. Nestled in the picturesque valley at the doorstep to Green Mountain National Forest on its east and the shadow of 3,855-foot Mount Equinox in the Taconic Range to the west, the village of Manchester has been described as the quintessential Vermont town. Its streetscape is lined with immaculately maintained green-shuttered, white clapboard houses. The tall white spire of the First Congressional Church—established in the late 1700s—and the building's Victorian architecture accent a row of historic structures along historic VT 7A as you enter the town from the south. The majestic Equinox Hotel, situated just across the street from the church, figures into the history of Manchester's tourist industry, which began to blossom in the mid-1800s. Fly anglers soon discovered the area and descended on the Batten Kill, which flows through the village.

Even before you get on the water, which is a productive walking distance from the Equinox, take time to visit the American Museum of Fly Fishing, located on VT 7A between Manchester Village and Manchester Center. Manchester is home to the Orvis company, and the awesome Orvis flagship store complex sits right next to the museum. An expansive facility, complete with trout ponds, a fly-fishing school, and a two-story retail store, it's a popular destination for fly-fishing visitors.

The spire at First Congressional Church sits across from the Equinox Hotel, a favorite vacation place of President Abraham Lincoln's widow, Mary.

The American Museum of Fly Fishing is a must-see during a fly-fishing visit to the Batten Kill. It sits right across from the Orvis flagship store.

Batten Kill fishing reports, including hatch updates, are typically posted inside the Orvis flagship store along VT 7A. Plenty of flies fill the bins in the fishing section. Any last-minute items such as leaders or lines and even Batten Kill angling advice are available.

Vermont Batten Kill brown trout are strong, colorful, and feisty. Browns inhabit water throughout Batten Kill's 59-mile length.

HEADWATERS ("EAST BRANCH") AND WEST BRANCH TO MANCHESTER

While brook trout can be encountered many miles downstream, well into New York State's reaches, the headwaters and the West Branch provide decent brookie fishing. Some refer to the headwaters of the Batten Kill as the "East Branch" yet USGS topography maps don't make that distinction.

A common area to try not far from Manchester Center is located off Dufresne Pond Road, a 10-minute drive from Manchester Center, via VT 11 West/VT 30 North. I've been told you can also access the stream a couple of miles from Manchester Center at a VT 7A bridge crossing on the way to East Dorset (near the Chantecleer Restaurant). I've never fished that stretch myself, and don't know much about its potential.

Another headwater access reach I didn't discover myself, but was recommended to me several years ago, is located just a few miles from the traffic circle in Manchester Center. Take VT 7A about 4 miles north from the traffic circle. Just past a VT 7A bridge crossing, a short gravel road leads to a small bridge crossing. You can park just before the bridge and try upstream. A short distance upstream, Mad Tom Brook on the right joins the much smaller Batten Kill on the left, just after Mad Tom Brook flows under the US 7 bridge crossing. I've been told some small pools and runs hold brook trout, but on the one occasion I fished it during late summer, I came up blank.

If you're up that far exploring the headwaters, stop by an easy trail access area located upstream. Continue driving north on VT 7A to Mad Tom Road, located directly across from Dorset Country Store. Turn right on Mad Tom Road and continue about 0.5 mile to a formal Mad Tom Brook parking area and trailhead, marked with a large "P" sign. You'll encounter more hikers than anglers along the trail that follows the brook. Back in 2011, Tropical Storm Irene ravaged the little brook, as well as trail sections. You'll encounter an informational kiosk at the parking area. In the fall of 2023 I performed a DNA sampling at the parking

Mad Tom Brook is a small Batten Kill tributary. Small brook and brown trout inhabit its pocketwater pools. You'll use hip boots or even knee-high boots when fishing it late summer.

area stretch. DNA sequencing showed presence of both brook and brown trout, with, not surprisingly, 100 percent base-pair matches. Brook trout DNA predominated the sample.

Cast small bead-head patterns up into the small pools for feisty bookies and an occasional little wild brown trout. I use my 6-foot, 3-weight cane rod while fishing Mad Tom and Batten Kill headwaters. Hip boots serve me well while stone-hopping the little brook.

The soft brushy West Branch can be accessed from VT 30 North out of Manchester. A parking turnout is on the right, 1.7 miles from the traffic circle in Manchester. Make your way down to the tiny stream with a bit of bushwhacking. You can get into the stream at the Mount Aeolus Road bridge crossing, also pretty small water. I rarely fish the West Branch, but I occasionally do so in late season, wet wading along the riffles. The fish are usually pretty small, but once in a while you'll net a dollar bill–size brookie.

MANCHESTER TO ARLINGTON REACHES

Ample access is available on reaches between Manchester and Arlington, particularly at bridge crossings. One "starter" reach is located on Union Street Fork, about a mile from its intersection with Main Street, which runs directly in front of the Equinox Hotel. Park down near the bridge and sample the stream with small dry flies and small-size streamers. My typical rod of choice while casting along the stretch is a Becker 7-foot, 4-weight cane rod.

Tricorythodes spinner falls happen midmorning on Batten Kill's flats around Sunderland.

The Batten Kill flats upstream from Arlington bring both brook and brown trout to the surface during August into September when the *Tricorythodes* emerge. Its best to get to the water early in the morning and wait patiently for the show to begin.

Another section relatively close to Manchester worth a look is the slow-gradient, slow-water reach located a couple of miles downstream at a small bridge crossing on Richville Road. From the Equinox Hotel on Main Street, drive less than 0.5 mile south on VT 7A to a left on River Road. Follow River Road about 2 miles to Richville Road and its bridge crossing. The Batten Kill's low-gradient flow, sometimes difficult to wade, harbors brown trout along the flats that often show themselves during *Tricorythodes* spinner falls during August into September. During late season, I've had some success plopping deer hair ant pattens tight to the bank deadfalls.

On its way toward Arlington, the river maintains its low gradient sinuous nature, flowing in and out of areas difficult to access. Bridge crossings are your best bet to get into water, one located off Sunderland Road. I've spent more time gazing at the flats located at bridge crossings, looking for rising fish early in the morning, than actually fishing (which I should have been doing).

One stretch in Sunderland (just upstream from Arlington) that I have devoted significant time to fishing and exploring throughout the season, but mostly when the *Tricorythodes* are in the air, is the Hill Farm Riverside Conservation reach on Hill Farm Road. It is located about 4.5 miles south of Manchester on VT 7A and about 3.5 miles north of the VT 7A/VT 313 intersection in Arlington. Park at the bridge crossing, and walk several hundred feet down the wide maintained trail that begins directly in front of the parking turnout to points where you can get into the river.

Mixed in with many successful Trico outings on the Vermont Batten Kill's Hill Farm reach, I've also experienced the agony of defeat. Such a situation occurred a couple of years ago. My typical Trico hatch success on that particular river reach in Sunderland was generally limited to small yet very colorful wild brown trout and brook trout, although I know that section supports sizable brown trout that are sometimes netted earlier in the season. However, I spotted an obviously larger trout feeding not far off the streambank in a side current, behind a deadfall tree branch that projected a few feet out into the secondary flow.

An audience on social media waited for my almost daily posts involving what would turn out to be some dozen outings in my quest for a trout I named "Themis." While I was able to entice a couple dollar bill–size brown trout over the course of a month on that stretch, Themis eluded me because of my failure to achieve a drag-free float over its feeding lane. A current change at the deadfall sheltering the trout was the culprit. I tried every approach imaginable, even a drastic total upstream cast to the fish that I rarely attempt while fishing Trico hatches. Such contests occur every season during Batten Kill hatches; in this case Themis won.

ARLINGTON TO NEW YORK BORDER REACHES

Batten Kill's mileage flowing downstream from Arlington provides very good fishing for wild brown trout, with a few brook trout in the mix. Ample public access is available along the 6.7 miles of VT 313 between the VT 313/VT 7A intersection and the New York state border. The Keelan reach, a favorite stretch of mine, is located about 0.6 mile from the VT 313/VT 7A intersection. Turn left on River Road. You'll see a metal gate, with a small "Keelan House" sign attached, on the left side of the road just before the bridge crossing. Trails that cross quaint footbridges lead to the river. Popular with both dog walkers and anglers, signage clearly indicates that hiking and fishing are allowed on the private land owned by Arlington resident Don Keelan, an active community member. Keelan once operated a bed-and-breakfast on the property where he still resides.

Red Quills (male Hendricksons) blanket many Batten Kill reaches in both Vermont and New York. It's probably the most anticipated hatch on the river, by fish as well as anglers.

Incidentally, on your drive west on VT 313 from Arlington, you'll pass the home artist John "Jack" Atherton (1900–1952) built in 1947. It's located on the left side of the road about 3 miles from the VT 313/VT 7A intersection, about 0.2 mile before you reach the Wayside Country Store. Atherton is best known for his dry fly series crafted with influences from the Impressionist school of artists—Monet and Renoir. His now-classic *Fly and the Fish* (1951) detailed his theories. His dry flies were designed for fishing hatches most anywhere, especially on the Batten Kill.

Even during Atherton's years fishing the Arlington stretches, the Hendrickson hatch *Ephemerella subvaria* was the mayfly the Batten Kill was best known for. Its annual appearance summoned fly fishers from all around to fish the river. In her memoir *The Fly Fisher and the River* (1995, 2015), Jack's wife, Maxine ("Max"), described driving up the road into Arlington and encountering a Hendrickson hatch. When that occurred, she would turn the car around and head back to the house, alerting Jack in his art studio.

While Atherton's pinkish body Number Four dry fly has always been my favorite on both the Arlington and New York reaches, when Hendricksons are emerging, other contemporary Hendrickson designs are also very effective during that hatch, which typically occurs from mid-April into May. My friends Henry Ramsay and Tom Miller tie patterns that are among the best.

Continuing west on VT 313, you'll pass Wayside Country Store (a good place to get a sandwich or cup of coffee). A mile later you'll encounter the quaint red covered bridge (on Covered Bridge Road). Illustrator Norman Rockwell once lived in the large white house at the end of the road, during the same time Atherton lived in the area. A popular formal fishing access parking lot is located about 3 miles past Wayside Country Store, about 0.5 mile before reaching the New York state border. If you're fishing this close to the Vermont–New York border, it makes sense to try the variety of water available on the lower river. It makes no sense to discuss Batten Kill without describing New York's share of the river.

NEW YORK STATE REACHES

You'll find extensive coverage and details of New York's 32-mile-long section of the Batten Kill in my *Fly Fishing Guide to New York State* (2023). However, I'll provide a few highlights here in the event anglers find themselves curious about what the lower river provides.

The first 4 miles in New York's Batten Kill, from the Vermont border downstream to the quaint Eagleville Covered Bridge, is regulated as year-round catch-and-release (C&R) water. This is by far the most popular fly-fishing section on the river, but also the most popular mileage among recreational kayak, canoe, and inner tube enthusiasts. The multiple-use concept is alive and well on the river—sometimes a bit *too* alive and well for many fly fishers, including this one.

During the late 1970s, when I first explored this river, fishing was rarely interrupted by recreational floaters. Friendly canoers would float by now and then, with a friendly wave as they drifted by. That all changed when the first commercial inner tube establishments colonized the stream. The "industry" grew, expanding into Vermont. Then came the kayak and canoe renters.

During the hot summer months, the NY 313 rest stop turnout, located only a short distance from the Vermont border along the C&R stretch, swarms with swimmers and floaters. The only way I can describe just how bad it has become is to liken the river to a theme park. Commercial canoe and tube rental businesses cart dozens of floaters to the convenient put-in at the rest stop, creating much angst among anglers. Some anglers never return.

One of my favorite New York C&R stretches during early spring and early fall is the Grocery Pool and the glide below it. I like to slip into the river here very early in the morning before the flotillas begin, and then later in the evening. Just past the NY 313 rest stop you'll see a large white house on the road. The Grocery Pool, named for the old red grocery store that existed nearby in the 1800s, is located several hundred feet behind the large white house and cornfield along the road. Park at the NY 313 rest stop turnout and walk downstream along a river path or park at the NY 313 bridge, located a mile or so from the Vermont border, and fish upstream to the pool. Before heading down to the Grocery Pool, try Dutchman's Hole, just upstream from the rest area. A small bridge spans the river, marking the pool and the glide below it. It's best known for the then–state record 31-inch, 12½-pound brown trout that area resident Roy Brown took on May 11, 1923. Brown also took a 6-pounder out of Dutchman's in 1942.

I've not repeated Brown's accomplishments on Dutchman's, but I've netted some big browns out of the Grocery Pool, a hemlock-shaded ledge-rock hole at the base of a steep mountain, well away from the main road. The pool is known for large trout. I twice hooked and lost a big brown I named Tethys (for the Greek Titan of fresh water). I finally netted the fish on an old Catskill-style dry fly called the Firehole No. 1 after pursuing the trout for a good month, before sunrise on some days and until last light on others. I've taken some nice fish casting and skittering dry flies along the heavily shaded hemlock banks downstream from the Grocery Pool, working my way down to the NY 313 bridge. After sampling that area, try Camden Creek, located off Hickory Hill Road. (The road begins next to the NY 313 bridge.)

Camden Creek, a Batten Kill tributary, provides public fishing rights that extend both above and below the bridge in the typically crystal-clear stream. I've had great fun fishing upstream with my 7-foot, 4-weight cane rod, catching the wild brookies on dries. However, the creek's mouth, where it pours into the Batten Kill, has always been a favorite spot. Located within walking distance of a fishing cabin I once owned 5 minutes away on Perry Hill Road, I call the stretch Camden Creek Run. It's relatively secluded and well shaded by overhanging hemlocks.

While it's best to avoid Camden Creek Run on the Batten Kill during hot summer days because of heavy inner tube and kayak traffic, fishing can be good in early May and again beginning in September. I've encountered everything from unbelievable swarms of American Grannom caddisfly flights and heavy Hendrickson flotillas in May to Trico dances in August. However, I've done best here by swinging small bucktails fished across and downstream through the moderate flowing current. A classic Batten Kill bucktail called the Shushan Postmaster works well. Lew Oatman (1902–1958) created the fly many years ago for Al Prindle, the town's beloved postmaster.

Oatman, who was born and raised on the Batten Kill in Greenwich, and who later acquired a summer place on Roberson Road in Shushan, is considered the father of baitfish fly imitations. The house he once occupied on Roberson Road is within walking distance of the next important pool, Spring Hole, within sight of Camden Creek Run.

The Spring Hole pool is located upstream from the CR 61 bridge (locally called Buffum's Bridge). A popular gathering place called the Tackle Box, once frequented by anglers, including Lew Oatman, was located on the bank at Buffum's Bridge. The small cabin that housed it is still there, although it's no longer a tackle shop. Lee Wulff lived in Shushan from 1941 to 1962—and once jumped off Buffum's Bridge with his waders to prove that he wouldn't drown wearing them.

Expect a lot of company at the Spring Hole, one of the most popular fly-fishing pools on the New York Batten Kill. The banks and parking lot are also heavily used by recreational kayakers, tubers, and swimmers. A Vermont canoe rental outfit often parks at the bridgeside angling access lot, ferrying clients. Fishing here during summer requires getting into the river at daybreak and getting back out by early morning. Avoid it altogether on summer weekends, especially holiday weekends.

The river has additional C&R stretches downstream from the Spring Hole. To reach them, drive back up CR 61 to NY 313 and turn right (southwest). About 0.25 mile down NY 313, a sign marks an angler parking turnout. A footpath leads to the river and a deep pool. I used to fish this area extensively 30 years ago, but lately I've passed it by, instead favoring a beautiful dry-fly stretch located a bit farther downstream; park at a turnout at the end of the guardrail 0.5 mile down NY 313 from the parking area just described. The stretch I enjoy fishing at dusk is located several hundred feet downstream from this turnout. The relatively shallow and placid glide is easy to wade and perfect for fishing small flies over risers. The C&R section ends at the Eagleville Covered Bridge, some distance below Wulff's beat, the next popular piece of water if you continue to fish downstream. Lee Wulff once lived in the red dwelling in view on the hillside along the river.

While the Batten Kill's C&R water is the most popular, fly anglers shouldn't ignore other stretches downstream. Pook's Bridge, located on CR 64 in Shushan and named for Charles "Pook" Gilchrist, who owned land next to the stream can be productive both upstream and downstream of the bridge. The Georgi on the Battenkill Museum and Park grounds, located on Adams Lane off CR 61, in the tiny hamlet of Shushan next to Yushak's Market, is another nice spot. The lower end of the river along the Georgi grounds flows under a high CR 61 bridge and out of sight. A large pool is located on the downstream side of the bridge crossing. An old covered bridge (now a museum) is situated next to the modern bridge. Sample the pool and runs by parking at the covered bridge museum. A well-worn path that leads from the museum's parking area will get you into the water.

Few anglers fish the 3-plus remote miles of river that lead from the Georgi stretch to the Rexleigh Covered Bridge, but this isolated stretch produces some of the Batten Kill's largest brown trout. I access the water by parking at the formal (New York State Department of Environmental Conservation [NYSDEC]) parking turnout on Rexleigh Road next to the historic red covered bridge; reach Rexleigh Road by taking CR 61 west to NY 22. On lower water, wade and cast on the downstream side of the bridge. Trout hold along the far bank.

About a mile downstream from the Rexleigh bridge, the river flows under the NY 22 bridge. You can park at a NYSDEC lot next to the bridge. I prefer to fish upstream from the bridge to target fish that hold in the shaded areas along the north. Swinging small wet flies or bouncing caddisfly dry flies through and along the riffles is effective. Wading is easy upstream from the bridge, a perfect spot to introduce a youngster to the sport.

On the lower Batten Kill, miles of water remain seldom fished but have produced some huge fish. On April 8, 2011, Shushan angler J. T. Trainor took a 29.5-inch, 9-pound brown on a Rapala. During my early years on the Batten Kill, I frequented a reach of the lower river that I named Adams Pool, after Bill Adams, the fly angler who showed me the area in

1978. I first described Adams Pool in the November/December 2016 issue of *Eastern Fly Fishing* magazine, bringing it to the attention of others who had never fished it, and never knew I had assigned its name. Over many years, I alerted other anglers not familiar with the lower river that Adams Pool is a good area to encounter Hendrickson and caddis hatches. It's located next to a turnout on the left side of the road 5 miles west of the NY 22/NY 29 intersection. During one session, Adams Pool gave up a 19-inch wild brown that sucked in a dry fly I call the Batten Kill Badger during a Hendrickson hatch. During the late 1970s, the glide above the pool used to be great during late May, when trout rose persistently in the evenings to caddis skittering near the water surface. I'm not sure how well it fishes today.

Areas downstream toward Battenville, off Skellie Road, also hold trout. Pools and runs upstream and downstream from the CR 61 bridge crossing (off NY 29) are popular with fly fishers, especially during the Hendrickson hatch. Downstream in the Greenwich area, a pool repulsively called the Sewer Hole off Mill Street in Greenwich, also holds a few trout, but they rarely show themselves during hatches. It's best fished during early and late season.

HATCHES

While I most enjoy fishing the Trico and Hendrickson hatches, there are other emergences that bring trout to the surface. Another favorite of mine is the Grannom caddis (*Brachycentrus)* swarms during early May. The browns generally feed subsurface during the morning swarms, and Tom Baltz's CDC Grannom Pupa is very effective. Clinging mayfly species such as March Browns and Quill Gordons are generally absent or uncommon.

Blue-Winged Olives, Little Black Stonefly (*Taeniopteryx nivalis*), and Little Brown Stonefly (*Strophopteryx fasciata*) are on the water during early Hendrickson hatches. I've encountered hatches of Blue Quills (*Paraleptophlebia adoptiva*) during early May. Blue-Winged Olives of various genera continue to hatch any time from July into October. Sulphurs (*Ephemerella dorothea dorothea*) show up here and there along the river during late May and June. October caddis, often called the Great Brown Autumn Sedge (*Pycnopsyche*), appears later in the season, usually in the early morning or late in the evening, but not in swarms. Flying ants make major flights during the first humid day in September, bringing risers to the surface.

CDC Grannom Pupa *Tied by Tom Baltz*

Hendrickson CDC Thorax Dun *Tied by Henry Ramsay*

Female Rusty Spinner *Tied by Henry Ramsay*

As far as favorite hatches are concerned, local resident and longtime Batten Kill fly angler Tom Miller has his favorite. Tom is well known for his sought-after, well-tied flies that work well on any Batten Kill stretch. Having fished the river for decades, Tom understands the hatches well. While many fly anglers look forward to the Hendrickson emergence, Tom is much more excited about another mayfly that appears on Batten Kill stretches—Blue-Winged Olives.

"Far and away my favorite hatches on the 'Kill are of the Olives, specifically those of the later season, where #24 and #26 flies are the norm. The weather in October is pleasant and the river is cool and clear; I can chase grouse in the morning and BWOs in the afternoon and find fish just about everywhere I look for them. The trout have been accustomed to surface feeding for 5 months, and are selectively looking upwards for their meals; as a dry fly man I would have it no other way."

Hendricksons are important to many anglers, and you should have an ample supply of effective patterns for that bug during April and into May; but as many of us have noticed, you'll often encounter lots of duns on the water with few rising trout. Tom shared additional thoughts:

"Hendricksons? Nope," Tom commented. "Water levels/temps too variable early in the season to have reliable surface activity. Sulphurs—great hatch but confined to last hour of daylight most days. Isos [*Isonychia*] too spotty to provide consistent sport. The Yellow Quill (*vitreous*) is, in my opinion, the prettiest mayfly on the river, but not nearly plentiful enough. The big drakes—March Browns, Gray Foxes, and Cahills are present, but practically nonexistent. Tricos are fun and challenging, but let's face it—mostly small fish that even a 2-weight and 18-foot leaders can't apologize for. Caddis, Stoneflies, and Midges are out, as well as terrestrials. Blue Quills are good, and provide decent fishing with their

Batten Kill Flats *Tied by Tom Miller*

While many have not heard of this awesome parachute Batten Kill Flats fly, it is one of the most effective fished on Batten Kill during the Hendrickson hatches that come off the water in April into May. Tom Miller said the pattern is "a great representation of the myriad rust-colored spinners commonly found anywhere there are mayflies." Stripped Rhode Island Red hackle quill is used for the body. Tom wraps parachute dun hackle larger than usual, by at least one hook size.

Sailwing Hendrickson *Tied by Tom Miller*

"This fly was devised by George Schlotter to imitate the more popular mayflies that are commonly taken by trout in their dun (winged) stage, and is a great producer on the Batten Kill." —Tom Miller

Blue-Winged Olive *Tied by Tom Miller*

Tom's BWO version of Vince Marinaro's classic Cut Wing Thorax Dun

Vermont Caddis (aka Hares' Ear Caddis) *Created and tied by George Schlotter*

George Schlotter's former Angler's Nook fly shop in Shushan once served as a first stop for any fly fisher heading into the Batten Kill. His Vermont Caddis was among the dry flies that hopeful anglers were after before heading out on water. It's roughly tied in small hook sizes, heavily overwrapped with grizzly hackle along its scraggly body.

multiple broods, but pale in comparison to the Olives."

However, the favorite hatch of anglers such as experienced Batten Kill fly-fishing guide and long-term local resident Scott Sztorc is indeed the Hendrickson. Scott and Tom are good friends, and both respect alternative views on Batten Kill tactics. The three of us gathered on the banks of the 'Kill one April day, just across the border from Vermont on New York's reach. We enjoyed a friendly discussion about flies and preferences. I asked Scott more recently about Batten Kill hatch preferences:

"I would definitely say my favorite hatch is the Hendrickson. The reason I love this hatch is that it's a great indicator that spring has finally arrived, and there are so many beautiful fish rising to the occasion. And of course, like so many other fly anglers, I start thinking and dreaming of this hatch in January while awaiting another spring season on the Batten Kill to arrive. My favorite flies to fish during this hatch are the Conover and the size 16 spinner pattern."

The Conover, of Catskill origin, is a very old classic dry fly. I watched in amazement that April day when Scott enticed and landed more than one beautiful Batten Kill brown trout on that oldie but goodie dry fly. I watched Scott from my vantage point on the riverbank, while he laid out precision casts to risers as Hendrickson duns were emerging.

Green River

Equinox Mountain's southeast slopes drain easterly into the Batten Kill valley. The northwest slopes also contribute water to the Batten Kill by draining into the 10-mile-long Green River, a small tributary born near Beartown. It flows southwesterly through Sandgate, picking up additional little brooks before wedding the Batten Kill in West Arlington. The Green parallels Sandgate Road before eventually crossing VT 313 at West Arlington, some 13 miles from Manchester Center, where it dumps into the Batten Kill. Sandgate Road is located about 3.8 miles west of Arlington and about 2.8 miles east of the Vermont–New York border. While most of the stream flows through inaccessible private lands, you can get into the water at a couple of points.

From the VT 313/Sandgate Road intersection, drive north 0.2 mile on Sandgate to Green River Drive. The gravel road connects back to Sandgate Road via a 1-mile-long loop. The road sign is marked as private, but I've never been asked to leave the stream that flows next to the road. What you don't want to do is park along the road shoulder anywhere across from the small private dwellings. It's best to find a place to park more toward the section where it rejoins Sandgate Road (1.1 miles north of the VT 313/Sandgate Road intersection) then walk to areas where you can get into the water, using common sense to avoid encroaching on the private dwellings. You won't see posted signs, but always respect private property. VF&WD fisheries staff regularly survey Green River at a site on Green River Road.

Green River, a small Batten Kill tributary, is actually more of a small creek, or brook, that joins the Batten Kill in West Arlington.

The easiest access on the upper reach is located about 1.8 miles north of the VT 313/Sandgate Road intersection at an obvious road shoulder parking area popular with kids looking to get into the cold water during summer's heat. It's located next to Southeast Corners Road. Make sure you don't block a fire department hydrant on the paved shoulder. I enjoy fishing downstream from this point, through the riffles, again respecting private property. The lower reach from its confluence with Batten Kill upstream of VT 313 is perhaps the most fished, and the most evaluated by VF&WD. Data taken annually shows a good population of both brown and brook trout, mostly smaller and young-of-the-year size ranges.

Anglers who fish this stream for the first time will immediately wonder why the Green is labeled a river. It's really no more than a small creek or brook. It may not fit the definition of a river, but Green River actually looks green, a shade created by its subsurface greenish schist. The best way to fish the stream is to wade carefully upstream, hitting all likely spots. Given the very clear water, stealth is essential. Brook and a few small brown trout will smack small bead-head patterns or small dry flies such as Elk Hair Caddis.

Roaring Branch (of Kelly Stand/Arlington)

Roaring Branch (not to be confused with another Roaring Branch, over on the Walloomsac River near Bennington) is one of my favorite little brook trout waters. Born in the Green Mountain National Forest, it is formed from the waters of Branch Pond Brook and Alder Brook. The stream runs about 10 miles, mostly westerly, to East Arlington, where it picks up Warm Brook before eventually joining the Batten Kill.

If you happen to be sampling East Branch of the Upper Deerfield River headwaters or Black Brook in the Green Mountains along Stratton-Arlington Road (see page 241), you can get to the upper Roaring Branch by driving west on the seasonal gravel road that continues as Kelly Stand Road. Roaring Branch follows Kelly Stand Road as the boulder-strewn stream tumbles down the mountain toward East Arlington. I've always preferred fishing

One of sporting artist Adriano Manocchia's favorite brooks in the Arlington, Vermont, area is Roaring Branch. Adriano loves casting his small flies into the various pools and runs along the stream, especially during fall.

the upper reaches of Roaring Branch during late summer, wet wading through the little stream, casting small dry flies along tiny pools. Head for bridge #4, located about 4 miles west of the Stratton-Arlington–FR 71 intersection. A small parking turnout is located at the small bridge. I fish upstream from the bridge—water that is somewhat isolated, well away from the road, and tucked into the woods. Brook trout are small yet very colorful and fun to catch on the upper reach.

If you're fishing the Batten Kill in the Arlington area, head for Kelly Stand Road in East Arlington. Plenty of small parking turnouts and bridge crossings will get you into the water as you drive up the mountain along the gravel road that can be a bit rocky and rough at times. Fish along the stream while heading up to the headwaters (bridge #4 crossing is located about 5.8 miles from East Arlington, up Kelly Stand Road). Care should be taken while climbing through and around all the boulders and rocks while fishing.

Brook trout that inhabit Roaring Branch's uppermost reaches are small yet vibrant with color. Small dry flies smack patterns such as Elk Hair Caddis and Lime Sally Stonefly.

When Tropical Storm Irene hit the area in 2011, its massive downpour devastated the stream in the torrent crashing down the mountain valley. Kelly Stand Road was washed out and closed for an extended period but has since been repaired and is again passable. Sporting artist Adriano Manocchia, my fishing friend and neighbor in Cambridge, New York, often comments that the stream fished better for many brook trout prior to the washout, but we still enjoy fishing it.

Adriano prefers small bead-head patterns, which is probably a better choice; I stick with small dry flies, particularly John Checchia's Little Green Stone Fly or a Lime Sally Stone Fly. When I'm on the lowermost sections, near the confluence with the Batten Kill in Arlington (you can get into the water at a VT 7A bridge crossing, but respect private property), I'm likely to be stripping small size classic streamers, such as Lew Oatman's Golden Darter.

Golden Darter *Tied by Mike Valla*

In my book *Tying the Founding Flies* (2015), I mentioned Lew Oatman's now classic Golden Darter, how to tie it, and Oatman's fishing episode with the fly at the mouth of Roaring Branch at its confluence with the Batten Kill. In his *The Practical Fly Fisherman* (1953), A. J. McClane included correspondence with Oatman concerning the Golden Darter and the sizable brown trout Oatman netted on the streamer. Golden Darters still work well on the lower Roaring Branch itself, just upstream from the Batten Kill, where both brown and brook trout routinely smack the fly.

Branch Pond in the Green Mountain National Forest is more easily accessed than the more remote Bourn Pond. Canoes can be easily launched closed to the parking area on FR 10. The pond is located a short distance from the upper Deerfield River and Black Brook headwaters. VALERIE VALLA

Branch Pond, Bourn Pond, and Grout Pond

Anglers contemplating fishing the uppermost sections of the East Branch of the Deerfield River or Black Creek (see the Connecticut River Basin section) should consider sampling a few ponds in the same neighborhood, nestled in the Green Mountain National Forest. Kelly Stand Road and Stratton-Arlington Road are the gateways to two great brook trout ponds—Branch and Bourn—that await eager fly anglers hoping to net a large pond brookie. Branch Pond is relatively easy to access; Bourn Pond is more remote and requires a 5-mile round-trip trek through the forest.

If approaching from Arlington, Vermont, Kelly Stand Road, a seasonal gravel road out of East Arlington, will deliver you to the two pond trailheads and access points. Drive about 6.6 miles up Kelly Stand Road to a left turn onto FR 10. Driving another 2.4 miles on FR 10 will lead to a parking area and trailheads to both ponds.

Eager anglers can use an alternate route if approaching from areas near Manchester, Vermont. Take VT 30/11 easterly out of Manchester Center, and continue about 7 miles until VT 30 splits to the right. Turn right onto VT 30 South, then continue about 7 miles into Bondville. Just past the 7-Eleven store, turn right onto Stratton Mountain Road. Drive up the mountain about 5 miles, passing Stratton Mountain Ski Resort along the way, and turn right onto Mountain Road (just past a green-colored resort maintenance facility). Drive 4 miles and turn right onto West Jamaica Road; go 2 miles to Stratton-Arlington Road and turn right. You'll see signage to Grout Pond, a warmwater fishery.

Continue past the entry road to Grout Pond, another 4 miles or so to FR 10 on the right. (You'll pass the Upper East Branch of Deerfield River, Black Brook, and FR 71 on your way to FR 10.) If you are approaching from areas to the south, FR 71, mentioned in the Deerfield River section (see page 242), will get you in range of the pond trailheads. National

Bourn Pond's relative remoteness requires the use of a helicopter to stock brook trout. USDA Green Mountain National Forest staffer Scott Wixom is shown here planting fingerlings. COURTESY OF USDA GREEN MOUNTAIN NATIONAL FOREST

Grout Pond is popular with campers, day users, and anglers. The pond supports a variety of warmwater species. VALERIE VALLA

Geographic Green Mountain National Forest South map 748 clearly shows the ponds, parking areas, roads, and trails.

If your time is limited, try 35-acre Branch Pond. You'll need a canoe or float tube to effectively fish it. A short 0.3-mile portage from the parking area down a good trail gets your craft into the water. Primitive camping is allowed near the pond, but check the Forest Service office for regulations. Early spring and fall fishing are best. Hiking in 2.25 miles from the parking area to 48-acre Bourn Pond and camping would please any angler in search of a wilderness adventure. Primitive camping is allowed, but make sure to check the Forest Service regulations. An informational kiosk with a regional map is located at the Bourn Pond trailhead.

Grout Pond in the Green Mountain National Forest is easily located on the Stratton-Arlington Road, where signage gives clear directions to the Grout Pond Recreation Area. VALERIE VALLA

Fruit Cocktail *Tied by Dave Whitlock*

Both ponds are annually stocked with brook trout via helicopter. USDA Green Mountain National Forest biological technician Scott Wixom told me that the number of fingerlings stocked fluctuates based on the availability of fish. Bourn pond is typically stocked with 8,000 fish; Branch gets approximately 2,600. Branch also gets about 1,000 yearling brook trout in the 8- to 10-inch range via hatchery trucks near the boat access point. Scott also said many of those yearlings may not be making it from one lobe of the pond to the other because of the narrow, shallow channel separating the two lobes.

Grout Pond Recreation Area has 11 first-come, first-served campsites along the pond shore, 7 of which can be reached by boat; the others are walk-in sites. All sites have picnic tables, and a hand pump for drinking water is available at the parking area, along with a vault toilet. There's no fee to enjoy the facilities along Grout Pond.

The pond offers good sport for smallmouth bass, bluegills, and pickerel. Motorized craft are not allowed on the water, but it's easy to launch rowboats and canoes. Fly anglers who own a float tube will have a great time paddling among the lily pads, popping bass bugs for smallmouth.

Walloomsac River

The 16.8-mile-long Walloomsac begins where South Stream and Jewett Brook join just south of Bennington. A complex of other area streams that drain much of the south-central portions of Bennington County contributes to its hydrology. Among the tributaries that flow out of the Green Mountains to the north and west of Bennington are Bolles Brook, Bickford Hollow Brook, City Stream, and Furnace Brook. Cold Spring and Stamford Stream also contribute water to the Walloomsac. Roaring Branch, its major tributary, originates in Woodford Hollow, at the junction of Bolles Brook and City Stream. After departing the Bennington area, the much larger river soon crosses into New York and enters Hoosick Junction, where it dumps into the Hoosic River. The combined waters eventually enter the Hudson River.

On its way to the Hoosic River, the Walloomsac flows past the historic Bennington Battle Monument. Opened in 1891, the 306-foot monument rises high above the town of Bennington, Vermont. At the towering obelisk's base, a bronze marker engraved with the words of

The Henry Covered Bridge reach is one of the most popular and easily accessible stretches on the Walloomsac in Vermont. The covered bridge was originally built in 1840 but replaced in 1989 due to the original bridge's deterioration. VALERIE VALLA

Brigadier General John Stark (1728–1822) reminds tourists of the historic encounter: "There they are, boys! We beat them today or Molly Stark sleeps a widow tonight!"

And with that somewhat grandiloquent declaration, Stark launched the American Revolution's Battle of Bennington on August 16, 1777. Stark and his 2,200 militiamen, a force that included members of the Green Mountain Boys, were victorious over British forces; Molly Stark slept just fine that night. However, the battle wasn't fought in Bennington at all but rather on a promontory overlooking the Walloomsac River a few miles across the state line in Walloomsac, New York. While New York can claim the battleground, the Walloomsac River is born in the Green Mountains of Vermont. Yet, like the famous Revolutionary War battle, the river is shared by both states. Some of its most interesting features, however, are found in Vermont.

The 306-foot Bennington Battle Monument rises high above the Walloomsac as the river makes it way past the town of Bennington on its course toward the New York border a few miles downstream.

The Walloomsac is a much different trout stream from the renowned Batten Kill, which flows from a neighboring watershed to the north. While in Vermont, the Batten Kill is managed as a wild trout fishery with emphasis on habitat enhancement, the Walloomsac depends heavily on stocked fish. Streams with relatively low wild trout populations that attract a large number of anglers generally receive high priority for fish stocking. Wild brown trout and brook trout persist in the Walloomsac's headwater tributaries, yet the state stocks even some of these little streams. Fish are reared at the hatchery, 2 miles from downtown Bennington on South Stream Road (a nice place to visit if you're in the area).

The larger stocked fish, both brown trout and rainbows, are found in the Special Regulations Trophy Trout section that runs from the New York border upstream to the former Vermont Tissue Plant Dam at Murphy Road in Bennington. A thousand 2-year-old trout—rainbows and browns—are typically stocked in May at several release points along the river. The rainbows are triploid fish, meaning they have three of each chromosome, unlike natural fish, which have two of each chromosome. Two-year-old triploid rainbows tend to grow faster than diploid fish because they are sterile (no energy spent on reproduction).

The brown trout stocked in the special-regulation section are regular diploid fish. Diploid rainbows and browns are also stocked upstream from the boundary of the trophy section.

Because the stocking release dates are well publicized, the trophy section gets hammered by local bait anglers during early May, especially downstream of the dam at the former Vermont Tissue Plant and a couple of miles downstream at the historic Henry Covered Bridge, a red-painted quintessential Vermont covered bridge located at the intersection of River, Murphy, and Harrington Roads.

The stream suffers considerable early-season pressure, but enough fish remain beyond midseason. Once the crowds dissipate, it's somewhat easier to find elbow room. Most of the big rainbows are caught by the time the regular trout season ends on October 31. However, the trophy section, like most all Vermont streams, remains open all year to fly anglers.

The big triploid rainbows are not particularly fussy, and nearly any colorful streamer or bucktail pattern, such as an Edson Tiger, Little Brook Trout, or Mickey Finn, will entice fish. Several riffles and runs downstream from the Henry Bridge are productive, but expect plenty of company along this stretch of river, especially after the early-May stockings. The predominant mayfly is the little Sulphur (*Ephemerella dorothea dorothea*), a hatch that appears in late May or early June. The river produces few Hendricksons (*Ephemerella subvaria*), emerging in late April or early May, but in nowhere near the numbers encountered on the neighboring Batten Kill. Pheasant Tail Nymphs work well dead-drifted along the runs. You can park at the Henry Bridge or at a turnout located along River Road a short distance downstream.

The low-gradient, slow-moving river below the Henry Bridge is largely inaccessible from the point where it turns away from River Road, unless you're inclined to float it. However, the riffle where tiny Cold Spring Brook empties into the river, a few miles downstream from the Henry Bridge, is a nice easy-access location.

From the Henry Bridge, drive 2 miles north along Harrington Road to a turnout on the left side of the road just before Harrington joins VT 67. I've always called this section Cold Spring Brook Run. It's located only a mile upstream from the New York border. Incidentally, Cold Spring Brook itself, a trickle, is almost entirely inaccessible aside from a couple of stretches upstream a few miles along Rollin Road, where anglers can get into the water at the small bridges. The stream holds a few brook trout and small brown trout. Take note of the posted signs along the stream.

I latched into a couple of those triploid rainbows along Cold Spring Run a few years ago by surprise during a pleasant May morning when I was targeting holdover brown trout. New York stocks brown trout in the river a couple of miles downstream. Nothing was happening that morning on Batten Kill, so I wandered over to the New York section of the Walloomsac in the afternoon with the hope of avoiding a fishless day.

A few small *Baetis* mayflies were hatching on the New York water that day in May, but also striking out there compelled me to move upstream into Vermont, thinking a brown or two might be hanging off the mouth of Cold Spring Brook. Within minutes of hitting the water, I watched a deep arc bend my 5-weight rod. Because I was unaware that Vermont stocked that stretch with 15- to 16-inch rainbows, I assumed it was a big brown trout that struck my Little Brook Trout bucktail pattern.

While I'll admit some guilty pleasure in landing one of those triploid fish that day, I would have preferred getting into a holdover brown trout or, better yet, netting a few small wild brook trout and the occasional small wild brown that can be found in Walloomsac tributaries, especially the upper reaches of the Roaring Branch of the Walloomsac (see page

26). By the way, this Roaring Branch isn't the same stream that shares its name (a tributary of the Batten Kill) near Arlington, Vermont.

NEW YORK REACHES

When considering fishing the lowermost Vermont reaches, consider obtaining a short-term New York State fishing license. Much of this is covered in my book *Fly Fishing Guide to New York State* (2023). While it wasn't always the case before the water quality in this area of river improved, there's some decent New York water upstream from the Edward Cottrell Bridge (a mile downstream from Cold Spring Run) and at an area another mile downstream from that point at Caretaker Bridge along Caretaker Road.

For many years the Bennington Wastewater Treatment Plant (WWTP) had an adverse impact on the lower Walloomsac's water quality. The affected stretches included not only what is now designated as the Vermont Special Regulations Trophy Trout section but also most of the New York water. Stream studies conducted in 1984 pointed out the effects of sewage, such as persistent odor and river-bottom substrate covered with algae. Midges and worms make up 90 percent of the aquatic macroinvertebrate population in the river. However, water quality improved once the WWTP was upgraded in 1985. Based on macroinvertebrate and water-quality sampling, subsequent surveys have shown the river to be essentially unaffected.

Parking space for a couple of vehicles is located directly next to the stream at the Edward Cottrell Bridge. The railroad trestle run upstream from the bridge produces fish early in the season but falls off by summer. Bait casters typically set up shop at the parking area just after fish stockings in April.

Located only a couple of miles downstream from the Vermont border, the Caretaker Road stretch in New York is stocked with brown trout annually. VALERIE VALLA

Fishing is allowed along the Bennington Battlefield frontage. You can park next to Caretaker Bridge.

Little Brook Trout *Tied by Mike Valla*

Edson Tiger-light *Tied by Mike Valla*

Sulphur Quill Dun *Tied by Tom Miller, after A. K. Best pattern*

Another popular New York reach is located another mile downstream from the Cottrell Bridge, at Caretaker Bridge along Caretaker Road. General Stark and his army battled high on the hill above the Caretaker Road stretch. Fishing is allowed along the Bennington Battlefield State Historic Site river frontage. The run upstream at the railroad trestle crossing also produces trout, all hatchery fish. Your fly boxes should be filled with a variety of dry flies along with streamers and bucktails when fishing the Caretaker Road reach.

The Walloomsac River and its tributaries, such as the Roaring Branch, can provide fly anglers with a wide range of experiences. Walloomsac fishing, as well as the rich history of the Bennington area—from the historic Bennington Battle Monument and battlefields to the remote Glastenbury ghost town—provides an ample reason for anglers to wet a line in waters that flow in one of the nicest areas of southern Vermont.

Roaring Branch (of Walloomsac), City Stream, and Bolles Brook

You might find little pleasure in fishing the lower reaches of the Roaring Branch closer to Bennington that were heavily damaged by Tropical Storm Irene in 2011. Flood damage from the storm itself wasn't the primary culprit; rather it was the catastrophic destruction that resulted from knee-jerk decisions to deploy heavy earth-moving equipment in the streambed. The channelization activity ruined what was once trout habitat along several miles of the Roaring Branch.

However, other powerful hurricanes have damaged the Roaring Branch in the past, and the stream has healed over time. The New England hurricane of 1938 (often called the Long Island Express) brought millions of rocks and boulders down from the mountains and into

Bennington resident Morgan Boyd checks out City Stream at the Appalachian Trail footbridge crossing, upstream from its confluence with Roaring Branch.

the Walloomsac in the Bennington area, wiping out bridges in the process. Stream dredging and channelization by heavy equipment occurred, but not to the extent that happened after the more-recent catastrophic storm.

An upstream Roaring Branch reach worth a few casts in the Woodford Hollow area is located 2.5 miles east of the VT 9/VT 279 intersection in Bennington, at Harbour Road. Take VT 9 East to the intersection with Harbour Road, on the left. A small turnout is located about 0.1 mile up Harbour Road on the left, at a bridge that crosses City Stream. City Stream dumps into Bolles Brook at this point, creating the Roaring Branch (some maps call it Walloomsac Brook in this area).

Bolles Brook is one of a few Walloomsac tributaries that flow out of the Green Mountain National Forest. You can reach it by driving 2 miles up Harbour Road, located a few miles east of Bennington. VALERIE VALLA

Access to the junction of both streams is easy. The best access into City Stream is located at the Appalachian Trail/Long Trail access parking lot located off VT 9, 1.1 miles east of the VT 9–Harbour Road intersection. From the parking lot, take the short walk up the crushed stone trail past the privy to a handsome footbridge. Cross the footbridge and make your way down the slight slope into the stream. Another easy access point for fishing City Stream is located off Notch Road. From the Appalachian Trail/Long Trail parking area, continue east on VT 9 for 0.3 mile to a right turn on Notch Road. Then drive another 0.2 mile to an area where you can park and easily get into the water.

You'll no doubt pick up some colorful wild brookies and sometimes even a small wild brown trout in Roaring Branch and City Stream. For years I had assumed that the brook trout I caught in this area were all wild fish. However, the two streams are stocked with hatchery fish.

You can reach Bolles Brook's headwaters by driving 2 miles up Harbour Road from its intersection with VT 9. The pavement ends at a rather large "No Parking" sign. Drive beyond the sign, where the road enters the national forest and public access. The gravel road that enters the national forest at this point is better suited for off-road vehicles, but if you take it slow, cars with reasonable clearance can continue another 0.5 mile to a road gate and a turnout for parking. The gate is there to protect the fragile gravel road from that point on. Bolles Brook is located a couple hundred feet away. Within sight of the gate, some beautiful small pools and runs in the forest stream hold little wild brook trout. In the spring I fish it wearing hip boots; during summer it's a great wet-wade brook.

Upper stretches of Bolles Brook hold not only wild brookies but also much in the way of history and folklore. The Glastenbury ghost town, located along Bolles Brook about 2 miles upstream from the road gate, is renowned for giving some hikers (and anglers) the heebie-jeebies. During the 1800s the area supported a thriving charcoal-making industry until the mountains were eventually depleted of hardwood timber. What turned out to be an ill-fated resort designed to attract vacationers to the salubrious Green Mountains sprang

Pownal Tannery Hydroelectric Power Plant spills Hoosic River over its dam, which then flows through relatively short mileage until it crosses into New York. Multiple uses of the site have occurred over the years.

up here in the late 1800s but quickly vanished when Bolles Brook flooded and wiped out the rail lines that led to the community.

The Glastenbury area is now known for mysterious murders and disappearances, as well as for its historical intrigue. If fish aren't hitting along Bolles Brook, a hike up the gravel road takes curious anglers to the location of a once-thriving community. Few hints of its existence are now apparent, aside from a few old cellar holes buried here and there; its buildings and structures were swallowed up by the forest long ago.

Hoosic River

Beginning as a gristmill in the eighteenth century, it continued as cotton and woolen mills before being used as a tannery up until the 1930s. The dam site in North Pownal, Vermont, was revitalized as a hydroelectric plant by Hoosic River Hydro in 2017. Today, area residents and visitors alike enjoy the small Alan R. Strobridge Recreational Complex at the site, lined with a row of benches that overlook the Hoosic River. A kiosk at the little park reminds anglers of Annie Card, a 12-year-old mill worker photographed as the "Anemic Little Spinner in North Pownal Cotton Mill" by Lewis Hine in 1910. The image appeared on a US stamp, commemorating the first child labor laws.

Beside the kiosk is an interesting historical marker that describes how in the late 1700s local resident Margaret Krieger, accused of being a witch, was dropped into the freezing Hoosic River through an ice hole. If she floated, she was a witch. Of course she sank to the bottom and had to be rescued, living another several years, as the story is told.

Folklore about accused witches being dropped into the Hoosic River and historical facts about the misfortunes of children working in mills aren't the only things that bring people to this little riverside park to ponder the Hoosic River in North Pownal. Anglers of all persuasions come to the river in search of trout. While the Hoosic River along Vermont's short mileage can be difficult to manage by wade anglers at times, the bridge crossing in North Pownal at the dam area and several hundred feet downstream are good places to start.

While the Hoosic River flows some 70 miles from its beginnings in Massachusetts, fewer than 8 miles flow through Vermont before entering New York. While Vermont's mileage is relatively short, it's still worth a mention here. John Rogers, one of the most respected guides on the Vermont Hoosic section and other surrounding streams, enjoys the river and its environs.

"I love having the Hoosic in my quiver of southern Vermont options when planning a trip," John recently said. "I guide primarily the Batten Kill, the Walloomsac, and the Hoosic. For Vermont it's big water fishing; as you move down across the state line into New York, the abutting fields and distant mountain views open the sky up and it almost feels like out west. Consistently the biggest wild browns, I mean big (23-inch-plus fish), we put in the net come from the Hoosic, and I've never encountered another guide on my section on any of my floats over the past 7 years." John can be contacted at his website: www.vermontflyfishers.com.

Vermont stocks a number of 10-inch brown trout annually in Hoosic River in the Pownal area; sizable brown trout, as John Rogers mentioned, inhabit the river that's often mysterious to many. The river traverses New York, passing over other dams in Hoosic Falls, Johnsonville, Valley Falls, and Schaghticoke before eventually joining the Hudson River. Several small tributaries enter the river along its course.

The Vermont mileage isn't without challenges to fish, yet populations of both brown and rainbow trout that inhabit the river along the way make efforts well worth the time

The run directly downstream from the Dean Street bridge crossing in North Pownal next to the Alan R. Strobridge Recreational Complex is a nice section to sample when water levels allow for safe wading. Late May is a good time to sample that reach.

it sometimes takes to experience success. The majority of the river is low-gradient slow water, but brief riffles and runs interrupt its otherwise somewhat slow course. Water just downstream from the bridge crossing in North Pownal is one such run.

I like to fish from both sides of the river, depending on water levels. A footpath that follows the river can be found on the east side. From the recreational center, walk across the bridge to a point where a small light green utility building stands near to the road. The footpath is down the small slope. There are times when the river is best fished from that side and other occasions, again based on water levels, when the river can be fished from the recreational area side of the bridge. Rafts can be launched at the recreational area via a gravel road that leads to the river on the downstream side of the bridge. Launch your craft next to the river, but park back up on the side road. Some float anglers float downstream into New York water and take out near Indian Massacre Road, several miles downstream.

The deep slow-moving pool upstream from the bridge closer to the dam undoubtedly holds big fish. A perfect fly for that pool and other deep pools along the river (if you can get a cast out far enough) is noted Delaware River guide Ken Tutalo's Polar Sculpin. Providing water levels are safe, a run below the bridge crossing in North Pownal can accommodate wade angling. As is true on any large river, take caution when wading such areas. A good time to sample this reach is during mid-May and again in the fall.

If fishing the Hoosic along Vermont's North Pownal section, consider exploring nearby New York reaches. In my *Fly Fishing Guide to New York State* (2013), I describe other Hoosic River stretches flowing through New York in detail. I'll provide a few features here.

NEW YORK STATE REACHES

As is true in Vermont, wade-fishing on New York's reaches can often present challenges, but you can get into the water in a few stretches along NY 346. An interesting tributary stream that can provide good fishing for small wild rainbows is also in the neighborhood, just a few minutes from North Pownal.

A nice little tributary, Little Hoosic River (more of a small creek) is located just a few miles west of the Vermont–New York border. From the Dean Street bridge crossing, drive 1.6 miles westerly on VT 346 to New York. On your way to the Little Hoosic, try the run and deep pool on the Hoosic River located 1 mile west of the border. Park along NY 346 and carefully cross the railroad tracks. A short path leads to the nice plunging run and deep pool.

You'll cross the Little Hoosic bridge about 4 miles west of the New York border on NY 346. More details surrounding the Little Hoosic to its source are provided in my *Fly Fishing Guide to New York State*. Public fishing rights (PFR) stretches on Little Hoosic River provide ample access. Yet some sections are more conducive to fly angling than others. The headwaters in the Berlin area are narrow and very brushy and better suited to bait fishing. The better fly water is located from just south of Petersburgh to the confluence with the Hoosic River.

A good strategy is to begin a fishing day on the lowermost Little Hoosic PFR reach that's accessible several hundred feet from the intersection of NY 22 and NY 346 at North Petersburgh, just upstream from the confluence with the Hoosic River. It was once easier to access the water at the NY 346 bridge by parking on the road shoulder. However, a few years ago the shoulder parking was somewhat reduced when the bridge was refurbished. It's a bit tight, but you can still squeeze into the road shoulder at the bridge. I usually park

Little Hoosic river tributary, at the North Petersburgh area, is located only 4 miles from the Vermont–New York border, a short drive from North Pownal. Small yet wild rainbows inhabit the little stream that is managed as a wild trout water.

The Hoosic River at New York's Buskirk Bridge area. While the river appears more likened to bass water, wild rainbows and browns inhabit the lower reaches. VALERIE VALLA

Wild rainbows inhabit much of the Hoosic River in all three states it traverses. Browns predominate in some areas, while rainbows turn up in others.

back down near the NY 22–NY 346 intersection and walk to the bridge crossing and the path that leads into the stream.

About a mile of mostly secluded fishing in the section upstream from the bridge provides good trout habitat among the logjams and undercut banks. An old yellow-bodied, red-throated bucktail pattern called the Shushan Postmaster has produced well, fished through the sycamore tree deadfalls along the stream banks. Bypass the flat, shallow dead water you'll eventually encounter and continue upstream. There's a lovely boulder-strewn and shaded deeper section that can harbor some mighty nice trout.

The Little Hoosic joins the Hoosic River several hundred feet downstream from the NY 346 bridge at North Petersburgh. PFR water is available on the Little Hoosic all the way downstream to its confluence with the river. The slow-moving river at the confluence is popular with bait anglers, who frequently prop their rods up on forked sticks along the bank. Many sections are better fished with bait or lures than flies. Its appearance is more likened to a bass river. Most trout anglers would find it hard to believe that some nice wild rainbows come out of the river.

While the slow-flowing river doesn't appear like typical trout water, NYSDEC sampled the river years ago and found rainbows far downstream, even downstream from the Eagle Bridge section on NY 67. The rainbows and a scattering of brown trout tend to hang out in stretches where some degree of thermal refuge can be found. The low-gradient river warms significantly through the New York sections. Thermal refuge is found at locations where cold tributaries enter the stream. The Owl Kill weds the river at Eagle Bridge, providing some refuge. Smaller tributaries that enter the river help too. The Eagle Bridge reach is a popular fly-fishing location, and it's one of my favorite stretches because it's located only a couple of miles from my home in the town of Cambridge.

Polar Sculpin *Tied by Ken Tutalo*

Masked Avenger *Created and tied by Rich Strolis*

Vergennes Falls, in Vergennes, Vermont, is the final falls and dam that 112-mile-long Otter Creek, Vermont's longest river, encounters before it finally joins Lake Champlain, nearly 8 miles downstream. At one time or another at the fall's area, there's a chance to net almost any gamefish that inhabits Lake Champlain.

Lake Champlain Basin

Vermont's Lake Champlain Drainage Basin is vast. Significant water drainages that contribute to Lake Champlain originate in subbasins that include Southern Lake Champlain, Northern Lake Champlain, Missisquoi Bay, Lamoille, Winooski, and, collectively, Otter, Little Otter, and Lewis Creeks. Nearly half of Vermont's 9,614 square miles send water into Lake Champlain. Networks of small mountain brooks feed small creeks and rivers that end up in major waterways such as the Winooski, Missisquoi, and Lamoille Rivers. From their component basins, those three big river systems alone feed enormous water volumes to Champlain. It's no surprise that the enormity of the drainage basin supports a vast and variety-filled fishery. You'll enjoy casting for everything from small yet wild and beautiful brook trout that inhabit mountain brooks to big carp and bass and other species. While the emphasis of this book is focused on Vermont's inland fisheries, fly-fishing opportunities available on Lake Champlain itself, the mouths of tributaries, and reaches directly upstream from the mouths deserve mention. Lake Champlain's waters are located within the boundaries of Vermont, New York, and at its northern extent the Canadian province of Quebec. Named for Samuel de Champlain, the French explorer who set eyes on it in the 1600s, Lake Champlain supports an impressive diversity of gamefish that will strike flies.

Lake Champlain—Middle Section

Often overlooked by fly anglers, warmwater species such as bowfin and coldwater species such as lake trout can provide exciting experiences on Lake Champlain. Targeting both with fly tackle can prove frustrating yet enormously fun. For many years I've enjoyed casting for landlocked salmon and lake trout on nearby Lake George, something I described in *Fly Fishing Guide to New York State*. My primary interest was learning if fly-fishing techniques that proved effective for lake trout on Lake George were basically the same on Lake Champlain, a much larger waterbody. Lake trout on Lake George follow rainbow smelt into areas such as the mouth of West Brook, at the lake's south end. Smelt are not a primary forage on Lake Champlain.

I knew nothing about the mysterious bowfin, a surviving relic of the Triassic period, 250 million years ago. That all changed one April day while having lunch with Phil Monahan. "Have you ever caught a bowfin?" Phil asked. He scrolled through his phone images and showed me his, a specimen he caught while out with Drew Price up in the Ferrisburgh area on Champlain's backwaters. I had already planned a lake trout outing with Captain Price that following week. The plan was to get a laker in the net then try for bowfin later in the day.

Drew, who operates from his Master Class Angling, knows where the lake trout hang out that time of year, largely dependent on water temperatures, before they leave reasonably shallow water and head to the lake's depths. In addition to his sustained ability to get a wide variety of Champlain's fish species into the boat, he's known for his tremendous success in enticing lakers. "I've had great results casting Blane Chocklett's pink or white Game Changer patterns," Drew noted as we headed out on Lake Champlain close to Vermont's shoreline. Drew likes to throw an intermediate 8-weight sinking line off a 9-foot rod. While I was still getting my own tackle in order, Drew latched into a 28-inch laker—within 5 minutes!

Of course many different patterns are also effective lake trout enticers. One of the best is Rob Streeter's Laker Taker articulated streamer, which is very effective on Lake George, just 30 or so miles southwest of Lake Champlain. Rob ties the pattern in white shades also. My own feeling, having caught lake trout on many different waterbodies in New York, is that the most important factor is knowing where the fish are hanging out during spring (and sometimes fall). Drew's stance on water temperatures is very important, for sure.

After a couple of hours casting Game Changers for lake trout, Drew suggested we switch gears and head into his favorite backwater areas off Otter Creek for bowfin. Mid-April is a tough time to get into them, but Drew had a hunch we might locate a few—and we did. I likened the area to a Louisiana swamp, with hardwoods poking up through an area choked with fallen limbs and most of all with buttonbush in water only a few feet deep. We spotted a bowfin that I was unable to entice with a Mr. Bow Jiggles pattern, dropped on the fish's snout.

That bowfin darted away, but Drew spotted a second. "Get a short section of leader out, right over that fish!" It seemed impossible— in that dense cover, how could I possibly direct my fly through the buttonbush down to the bowfin? Somehow, armed with Drew's confidence I could manage it, and the fly made it down to the bowfin's mouth. The fish slammed it—fish on! But how in heck could I possibly pull it out from its buttonbush cover. A couple

Lake Champlain at Charlotte, not far from North Ferrisburgh. New York's Adirondack Mountains are in the background.

Drew Price caught this 28-inch laker during a mid-April outing with the author. Game Changer patterns are his flies of choice while targeting lake trout on Lake Champlain.

of minutes later, I had my first bowfin. "This is the earliest date in spring I've ever had one landed," Drew said as he shook my hand. That bowfin experience was exciting for sure, but plenty of other warmwater species ranging from gar to pike to carp and others can be had on Vermont's Lake Champlain backwaters and off tributary mouths near Vergennes. Lake Champlain's north area, near Missisquoi Bay, is also a fantastic area to get into multi warmwater species (see page 131).

Game Changer *Tied by Drew Price*

Drew Price prefers to cast Game Changer–style flies for lake trout on Lake Champlain during early spring. This is the actual fly he used the morning he latched into a fine 28-incher.

Some readers might already be aware of Drew Price's success not only in catching sizable lake trout and bowfin on Lake Champlain but also his accomplishments netting many other fish species that swim throughout Vermont's fisheries. Master Class Angling, Drew's guiding operation, is an apt choice of words to describe his abilities on water. In 2018 Drew completed an angling goal that's nothing less than remarkable. He completed Vermont's first-ever Master Anglers sweep by catching at least one of each trophy-size fish from a list of 33 eligible species. VF&WD sponsors the state's Master Angler Program. Drew's quest for this seemingly improbable goal began in 2010, when he entered his first trophy-size fish. By the end of that year, he had entered 12 species he took on a fly. Two years later he hit 24 with fly tackle. One species in particular, a pike-pickerel hybrid, proved to be a real challenge, but with the persistence Drew is known for, he caught that fish.

Mr. Bow Jiggles *Tied by Drew Price*

This is the actual fly the author used to catch his first bowfin on Lake Champlain.

The fly is fished by carefully lowering it straight down in front of the fish's snout.

Drew's Bugger *Tied by Drew Price*

Drew Price finds his bugger pattern, tied in various colors, effective for taking several Lake Champlain fish species. "I really like jig-style hooks for buggers because they are less likely to catch rocks and snags. I have found these 50-degree hooks combined with the tungsten jig bead acts like a conventional jig hook (which I have tied flies on for years)."

Captain Drew Price, who I call "Dean of Lake Champlain Fly Fishing," spotted a bowfin in a backwater area that flooded buttonbush. Here he drops a fly down to the snout of a bowfin that is holding in shallow water.

Mettawee River

Straddling two states, the Mettawee River is a charming trout stream. The stream provides great fly-fishing opportunity from its Vermont headwaters near Dorset all the way into New York. All 17 miles of its Vermont water is a wild trout fishery supporting brook, rainbow, and brown trout.

The Mettawee cuts across New York some 20 miles, eventually emptying its water into Lake Champlain near Whitehall. The lowermost New York section is a warmwater fishery. The upper New York public fishing stretches, the Granville area, are inhabited by stocked rainbow and brown trout, but you might hook into some holdover browns as well as a few wild fish.

The Vermont Mettawee hasn't received the same attention, or respect, given to its neighboring Batten Kill River. It's a much smaller stream, but its size belies the nice fish that swim in its quaint pools, many of which are hidden well away from roads. Unlike the Batten Kill, the Mettawee hasn't been the focus of major fly-fishing books or other exposure that would raise its profile. While the charming Mettawee is no secret to many, cane rod builders or talented fishing reel machinists haven't attached the river's name to those products. Major fly-fishing personalities haven't been attached to the stream's name (well, maybe one). Yet the Mettawee provides plenty for fly fishers willing to put time into exploring and understanding the stream in all its pleasant variety.

The Mettawee offers good water chemistry, with pH values exceeding 8.0, a great macroinvertebrate assemblage, and healthy stream-bred trout populations. Even better, the Mettawee flows through some of the nicest countryside you'll ever hope to visit—in both states. It all begins in its brook trout–filled headwaters.

The Mettawee River along River Road, not far upstream from the New York border, is one of the prettiest areas on the lower reach in Vermont. VALERIE VALLA

The little stream is born on the southern slopes of Dorset Mountain in Vermont, not far from its sister river, the Batten Kill, near Manchester—headquarters of the Orvis company. It flows down through the mountainside, edging the base of its namesake 2,800-feet mountain, The Mettawee. Brook trout inhabit the extreme headwaters, something retired USDA Green Mountain National Forest staffer Chris Alexopolous told me several years ago. Chris did some electrofishing in the headwaters back then and said, "The brook trout are not big in the extreme headwaters, but there's plenty of them."

Access is difficult along Upper and Lower Dorset Hollow Roads, the gateways to the uppermost headwater sections and the stream's birthplace. Both roads are located in and around Dorset, a small hamlet 6 miles west of Manchester. As is true with most of the Vermont Mettawee, access to the stream's headwaters is mainly found at the road bridge crossings. Much of the upper headwaters flow through Dorset Hollow, well away from the roads and easy access. However, there's a nice headwater access point along Peace Road in Dorset.

Peace Road joins VT 30 North in Dorset on the right just beyond the golf course. Cutler Memorial Forest, a nature trail area donated to the town in 1976 by Isabella Cutler, is located 0.5 mile up Peace Road near a bridge crossing the stream. An easy walk down the trail from the parking area will bring you to the stream.

Cutler Forest is a perfect place to bring along a non-fishing companion who can either walk the trails through the forest or sit and read a book on the small streamside bench under the forest canopy along the brook-size water. Bring your dog too.

The Cutler Memorial Forest access area provides fly fishing through small pools and runs. You'll likely catch small wild juvenile rainbows along this reach.
VALERIE VALLA

Fall is one of the nicest times of year to cast a line on the upper Mettawee River's headwaters in Dorset.
VALERIE VALLA

I like to enter the stream near the nature trail footbridge, fishing downstream 0.25 mile or so, through the pocketwater. It's one of my favorite stretches to fish in the fall season, using a 7-foot, 3-weight rod and picking up wild riffle rainbows on colorful small streamer flies. Brown and brook trout inhabit this area, but for some reason I always seem to catch only the little rainbows. A couple nice rainbows powdered a classic size 12 Shushan Postmaster streamer during my trip to the area last fall. The rainbows also hit a small bucktail I created years ago that I call a MV Mettawee Rainbow. Shushan Postmaster bucktails also work well.

MV Mettawee Rainbow *Tied by Mike Valla*

A flashy little bucktail pattern that Mettawee River rainbows really like, the fly is tied in sizes 12–14 on Mustad model 3665A hooks. The tail and throat are rusty red golden pheasant body feather fibers. Lion Brand Fishermen's Wool forms the body that's ribbed with embossed silver tinsel. Olive bucktail over pink bucktail creates the wing.

Departing Cutler Forest and the Dorset area and its quintessential New England charm of white clapboard, green-shuttered homes, the river's character begins to change. Eventually reaching East Rupert, the stream picks up water from small trickle tributaries then slows and meanders northwesterly along VT 30 through North Rupert then to Pawlet.

Shushan Postmaster *Tied by Mike Valla*

This bright old classic, created decades ago by Lew Oatman, brings plenty of strikes from Mettawee rainbows. Be sure to tie them in smaller sizes, down to 14s, for the juvenile rainbows that will strike them on the swing, fished through riffles.

The bucolic, wide-open Mettawee River valley beyond East Rupert along VT 30 North reminds anglers that this is farm country now. Freshly rolled hay bales dot the meadows, and cornfields border both the road and the stream. A couple of VT 30 North bridge crossings between East Rupert and Pawlet, a quaint town some 10 miles downstream from the Mettawee's source, provide access to the stream. The easiest access to the Mettawee is located 1.7 miles from East Rupert along VT 30 North at a formal VF&WD access parking area.

The stream is still very narrow upstream and downstream of the VF&WD fishing access parking area and appears, at first glance, unremarkable. Its misleading appearance would have some anglers thinking this section doesn't look like water to bother with. However, pools almost deep enough to swallow your waders are tucked away, perfect refuge for wild brown trout, brook trout, and rainbows.

There's plenty of fish food in this section of the Mettawee. Mayfly species in this area are mostly small Blue Quills (*Paraleptophlebia*) and Blue-Winged Olives (*Baetis tricaudatus* and *Acentrella turbida*). The Little Brown Sedge (*Lepidostoma*) and Grannom (*Brachycentrus americanus*) represent the majority of the caddis fly groups found here. Yellow Sally stoneflies (*Isoperla*) also inhabit the gravel runs. Fishing dry flies here is a fly-angler's nightmare, with overhanging tree branches and other obstructions not friendly to casting and getting a good fly drift. This isn't prime dry fly water. I fish this section of the Mettawee,

Mettawee at the VF&WD access parking area on VT 30 North. The stream here is small, perfect for shorter rod lengths. While much of the stream reach can be managed with hip boots in late season, some of the pools are very deep; waders are a better choice. VALERIE VALLA

which hugs VT 30N tightly, working tiny size 16–18 wet flies and nymphs, swinging them downstream through the gravel riffles. An Atherton #2 Hare's Ear nymph, tied in size 18, has worked well. Scrappy little wild rainbows willingly strike in the fastest current.

Leaving the VF&WD access area behind, my next stop when fishing the Vermont Mettawee is downstream in Pawlet, a quick 6 miles down VT 30N. You'll cross a couple small bridges that provide access to stretches worth a cast or two. You can park at the Mettawee Valley Community Center picnic and recreation area, located just short of the first bridge crossing, and walk to the stream. You can also get into the water at the second bridge crossing, a couple more miles down VT 30N at the intersection of Townslee Hill Road. But I usually skip these areas and head for the larger water in Pawlet.

Isonychia bicolor mayflies are one of several species that emerge from the lower Mettawee River in Vermont.

In Pawlet, take a left on Pawlet/Rupert Mountain Road at the Mach's General Store. Park at the Pawlet Library and walk down to the bridge. You'll experience pleasant fishing both upstream and downstream from the bridge. Slate Drakes (*Isonychia*) to the diminutive Blue-Winged Olives (*Baetis*) and Tricos are found in this stretch along with several caddis fly species. Trico emergences into early September can bring dimpling brown trout to the surface in the slow, deep pools. When trout are not on the rise, I like to fish downstream from the bridge using my Orvis 7.5-foot, 4-weight Superfine rod.

Mettawee River valley is among the most bucolic in Vermont. Straddled by dairy farms and other farm animal operations, the agricultural environment adds to the stream's charm.

Trout-inhabited Flower Brook enters the stream a short distance downstream, adding to the Mettawee's hydrology and width. You could spend the good part of a day exploring and fishing through a couple of miles of very secluded stretches where the Mettawee flows well away from VT 30N downstream from Pawlet.

The stream leaves the Pawlet area and pushes on another 7 or so miles to the New York border. The stretch along River Road, not far from New York, is one of my favorite Vermont Mettawee areas. River Road can be reached by driving along VT 30N 3 miles beyond Pawlet. There's a sign pointing to the Indian Hill Gallery just before River Road comes into view on your left. A mile down River Road will take you to Betts Bridge and a parking turnout that provides easy access to the stream.

You can fish a mile from the bridge, upstream or downstream. During the high-water early season, Silver Darter streamers, fished slow and deep, can bring strikes. During late spring I enjoy fishing upstream from the bridge, where the secluded Mettawee flows along a mountain base well removed from the road. I like to twitch an Adams dry fly in the small pools and dead-drift Atherton Hare's Ear nymphs in the pocketwater.

The River Road section gives up some nice brown trout every year to anglers who persevere, realizing that the Mettawee can be a tough river and not always willing to surrender its trophies. This wild trout fishery often discourages anglers who expect to land a trout on every outing. But the stream-bred trout are there, and netting one of them is an accomplishment.

I've experienced more than my share of blank days on Vermont Mettawee waters, but memories of wild trout that have come to my net persuade me to return to my favorite stretches season after season. Those in search of stocked trout waters, and much better chances of landing fish, will have to head to the New York stretches downstream. If you're fishing downstream that far on Vermont's stretches and want to sample the river's lowermost reaches, it makes sense to obtain at least a one-day New York State fishing license. Details of the New York reaches are found in my *Fly Fishing Guide to New York State* (2023). Highlights are discussed here.

THE NEW YORK METTAWEE SECTION

Leaving its 17 Vermont miles behind, the Mettawee crosses into Granville, New York, after flowing over impassible (and inaccessible) Button Falls, which keeps New York's stocked trout out of Vermont. The river picks up additional water from the Indian River, a meandering Vermont-born tributary. Fewer than 20 miles later, well beyond its impressive Class IV–V whitewater rapid sections around North Granville, the Mettawee finally dumps its water into the Champaign Canal and then finally into Lake Champaign south of the village of Whitehall.

The most popular New York Mettawee PFR trout water is located in Middle Granville, Truthville, and North Granville areas, where some 8,000 trout are stocked annually. The best way to reach these waters, and the formal fishing access parking areas, is to get to NY 22 at Granville, your gateway route to miles of stocked public fishing trout waters.

NY 22 is just a hop, skip, and jump from Vermont's River Road Mettawee section. At the end of River Road, take a left on VT 153 and drive 2 miles to West Pawlet. At the veteran's historical monument, located dead center in the road, take a right turn on Railroad Avenue, which almost immediately crosses into New York, becoming CR 29. You will reach NY 22 in 2.5 miles, where you'll turn right and head into Granville.

Granville is often referred to as "the colored slate capital of the world." Quarries scattered around the town mine green, gray, gray-black, purple, and red slates. The only working red slate quarry in the world is located here. You'll notice many roofs, along with other structures, displaying the colored slates the town is famous for. If the fishing is slow, or if you have time to do so, visit the Slate Valley Museum in Granville. Some wonderful displays tell the story of mid-1800s immigrants, arriving first from Wales, who came to the region with skills that helped transform the area into a booming industry that still thrives today.

The slate industry is apparent as you drive westerly along NY 22 en route to several NYSDEC stream access parking lots scattered along the river. One of a couple formal NYSDEC angler parking lots is located not far from where NY 22 merges with NY 22A at Middle Granville. You'll see a large red Bobcat farm equipment sign on your left, 1 mile north on NY 22A. Just beyond the sign and slate quarry tailings, the large NYSDEC angler parking lot comes into view on your right.

A footpath easement leads to the stream from the lot to the low-gradient Mettawee stretches. A pleasant streamside grassy path leads upstream, providing easy access to the riffles and pools. This is best fished in early spring, around the first couple of weeks of May, about the time when the Hendrickson hatch begins. The state stocks yearling rainbows in that section.

Other formal access points are located a short distance downstream from the Middle Granville NYSDEC parking lot. Drive another 1.5 miles north on NY 22A and then veer left onto CR 21 (you'll see Hilltop Slate on the left where CR 21 merges into NY 22A). In less than 0.5 mile, after the road crosses over the Mettawee, look for a small parking area on your left next to the bridge, on De Kalb Road. Because the New York Mettawee stretches are stocked, it's common to see local bait casters in action on the river for several days after the hatchery truck has made its annual appearance. They tend to fish from the banks close to the public angler-access parking areas.

A NYSDEC fishing parking lot in Truthville also provides easy access to the river. My largest holdover brown came out of this water several years ago during a late-May evening caddisfly flight—it's a stretch that sometimes prompts fish to rise at dusk. A 3.5-mile drive from the De Kalb Road fishing access parking lot gets you to the site; turn left out of the De Kalb Road parking lot, drive 2 miles to the end of De Kalb Road, and turn left onto

The Truthville reach is one of the more popular sections on New York's Mettawee. Elk hair Caddis dry flies work well skittered across the riffles. VALERIE VALLA

Truthville Road. Drive just over a mile on Truthville Road to Middleton Road and the CR 12 bridge that spans the Mettawee. The Truthville NYSDEC angler parking lot is located just 0.2 mile upstream from the bridge, on Middleton Road. You can also park at the bridge and walk down a steep trail to fish the pocketwater and pools downstream from the bridge.

I like to skitter caddisfly dry patterns over the riffles and runs in this area. Elk Hair Caddis patterns work well, but I also like caddisfly dry flies that have hackle in front of the wing to aid in skittering the fly across the water. Hendricksons emerge in early May. Small bucktails and streamers work well in this section when bugs are not in sight.

There's a final place anglers should visit when fishing the lower Mettawee River. Mettawee Falls plunges out of a gorge-like setting into a large, deep, swirling pool. You might run into a big rainbow or brown trout or smallmouth bass. There's also a good chance you won't net a single fish. You're more likely to run into bait anglers than fly fishers at Mettawee Falls. Expect swimmers and other recreationalists if you visit during the hot summer months. I most enjoy admiring Mettawee Falls in early May, about the time when the first signs of spring are evident in the shrubbery along the river. A massive rock slab dips into the pool, offering a great place to sit just after sunrise and have a cup of coffee while enjoying the scenery. For me, the place is more of a scenic destination than a fly-fishing destination.

Reach Mettawee Falls by taking NY 22 to the North Granville area from the NY 22/NY 22A junction in Middle Granville. In about 4.5 miles watch for the large white Calvary Baptist Church building on your right at Sheehan Road. Turn right onto Sheehan and drive 0.2 mile to Upper Turnpike Road, then turn left. The NYSDEC angler parking lot is 1 mile down Upper Turnpike Road (a NYSDEC angler parking sign on Upper Turnpike Road marks the location). A path leads down the hill to the falls.

In areas downstream from Mettawee Falls, all the way to the river's mouth at the Champlain Canal in Whitehall, the now turbid Mettawee flows at a low gradient, transforming

Mettawee Falls on New York's lower river reach is best fished during May. You might latch into anything from a big bass to a brown or rainbow trout in its swirling pool. VALERIE VALLA

into a warmwater fishery. By the time the Mettawee reaches Whitehall, the cold, once-crystalline waters found in the headwaters are long gone. By summer, most of the New York Mettawee warms too much for trout. Trout fishing can turn marginal in many of the stretches that provided good sport during May and early June. If fishing that far downstream, approaching the Whitehall area, consider the hop, skip, and jump that will get you over to the Poultney and Castleton Rivers. Check out Flower Brook, which flows into the Mettawee at Pawlet, before heading to Poultney River.

Flower Brook

Flower Brook flows southwesterly off the southern slope of Tinmouth Mountain. It later skirts the flanks of Fayette Mountain, Dutch Hill, and Mount Hope along its path. Along its short 7-mile journey from its headwaters to its mouth in Pawlet, Flower Brook first follows close to Little Village Road, where it picks up Purchase Brook, a tiny tributary. From there the stream roughly parallels Green Hill Road, passing a bridge crossing on Parker Road, then downstream to the bridge crossing at Lilly Hill Road, where it picks up Mountain Brook, another tiny tributary, then Beaver Brook a short distance later. Once the stream exits the Lilly Hill Road crossing, it vanishes from access until it flows past another crossing on VT 133, just 1.5 miles or so upstream from what has been called Flower Brook Gorge, at the old millsite next to Mach's General Store at the VT 30/VT 133 intersection. I've called Mach's General Store "the new and improved" Mach's; the once general store went through different owners before Gib Mach repurchased the place. It's now a favorite among visitors and locals for their outstanding market-like environment. The perfect place for your lunch break, they'll make you the best sandwich you've ever had.

Flower Brook at the Lilly Hill Road crossing, near the confluence of Mountain Brook. Flower Brook is a quaint little tributary that flows into the Mettawee River at Pawlet. VALERIE VALLA

Getting back to Flower Brook, the best place to experience the brook is by getting into the water at the bridge crossings. Access the mouth with Mettawee River by getting into the river at a bridge crossing just down the road from Mach's on School Street and fishing downstream.

The next easy-access bridge crossing is located 1.3 miles up VT 133 from the VT 133/VT 30 intersection across from Mach's. Keep on VT 133 where it veers left at its intersection with Danby-Pawlet Road. The stream can sometimes warm along this reach during summer, so be aware of water temperatures. You'll have to carefully get down the somewhat steep bank to get into the water.

Access is also available at the Lilly Hill Road crossing, another 3 miles upstream from the VT 133/Danby-Pawlet Road intersection. Take a left on Lilly Hill Road; the bridge is quickly in sight. Parking is not an issue on either side of the bridge, but park on the wide near-side shoulder. Footpath access (often obscured in late summer by goldenrod and other plant growth) is found across the road from the parking turnout. You'll have to trek over the small Mountain Brook, just upstream from old concrete bridge foundation remnants, with little problem. Once you get past the small brook, walk another 100 feet down remnants of a wide trail, then get into Flower Brook. You'll encounter a posted sign along the wide "trail." If you wish to continue down the trail, note the landowner's phone number on the sign. Access across the land is by permission only. From that point, you can fish upstream from the Lilly Hill Road bridge or downstream. Get down into Flower Brook before reaching the posted sign. By far my overall favorite reach from years ago, prior to Tropical Storm Irene in 2011, is located at the Parker Road crossing, a few miles upstream. After crossing the Lilly Hill Road bridge, take the quick left on Green Hill Road. Drive about 0.8 mile to a right on Parker; the bridge is in sight. Flower Brook flows through a nice stretch downstream from the bridge. It's perfect for your 7-foot, 3-weight rod. During late season, small dry flies will take fish.

Out of curiosity, in August 2024 I took a DNA sample in Flower Brook's extreme headwaters. The brook was cold and clear, just as recalled from past years. Results showed 100 percent base-pair matches for both brook and rainbow trout. Rainbow trout DNA values were very high, five times the amount of brook trout DNA detected flowing through that reach. No brown trout DNA was detected.

Despite being hammered by high rain events, not only Tropical Storm Irene in 2011 but also in recent years, the little stream has held on. I was surprised to see the abundance of rainbow trout DNA detected.

Poultney River

Poultney River is born in Rutland County between the Tinmouth and Spoon Mountains. The stream flows some 40 miles before it finally enters Lake Champlain at Whitehall. The lower Vermont river mileage, downstream from the Poultney–Fair Haven town line, forms the border with New York state in the towns of Whitehall and Fair Haven. Downstream from the Poultney–Fair Haven town line, Poultney receives water from the Castleton River (a trout stream) and a number of small brook and creek tributaries. The river widens and deepens along its path to Lake Champlain, at South Bay near Whitehall. On its way, a dozen miles upstream from its confluence with Lake Champlain waters, the river crashes over spectacular Carver Falls near Fair Haven. It's the highest major waterfalls in Vermont. Before the river ends its course, it first flows through a deep ravine at Carver Falls. Like many Lake Champlain tributaries, the Poultney River provides fly-fishing experiences

Poultney River at a midsection of the stream midway between the towns of Poultney and Middletown Springs, along the Slate Valley Trail. While water levels are higher during early spring, streamers fished though slow-moving pools can be effective. VALERIE VALLA

for warmwater species along its lowermost mileage and trout fishing at its middle section and headwaters.

UPPER RIVER: TROUT FISHERY

VF&WD doesn't stock brown trout in the Poultney River, although you'll encounter wild browns. However, around 2,000 brook trout are stocked annually from Middletown Springs downstream to York Street in Poultney. Middletown Springs is a small community located about 9 miles or so upstream from the VT 140/VT 30 intersection in Poultney (or about a 20-minute, 11-mile drive from the US 7/VT 140 intersection in Wallingford if you're fishing Otter Creek in Wallingford, to the east).

During the late 1800s, a hotel in the small community attracted visitors to its alleged healing medicinal springs. For a while in the town's history, the purportedly healing spring water was bottled and sold. Floods eventually buried the springs, which were uncovered in 1970 through volunteer efforts steered by the Middletown Springs Historical Society. A small semblance of the old springhouse now stands at Middletown Springs Park, where community users and visitors admire the informational kiosk that explains lots of neat history. The park, located off Burdock Avenue at the confluence of North Brook, is a nice place to stop for a lunch break. The streamside trail at the park is convenient for anglers to sample the small headwater reach. You'll want to use a 7-foot, 3-weight rod and small nymphs in this stream section. Larger water is available downstream.

From the VT 140/VT 133 intersection in Middletown Springs, drive north on VT 140 to a couple of bridge crossings where you can get into the water. The stream is mostly unposted but there are a few sections you'll have to watch for posted signs. My favorite reach is located about 4.6 miles downstream from Middletown Springs (about 3.8 miles

The first sign of flying aquatic bugs you'll encounter during early season is Little Black Stonefly (*Taeniopteryx nivalis*).

A.P. Black Beaver *Tied by Mike Valla*

Work slowly through slow-moving pools when Little Black Stonefly (*Taeniopteryx nivalis*) is on the water during March. Any small dark-shaded nymph in small sizes will work.

upstream from the VT 140/VT 30 intersection in Poultney). A section of the Slate Valley Trail System crosses VT 140 near the intersection with Town Farm Road. You'll see a small "Trail System" sign posted at a gravel road that leads down to a small bridge trail crossing. Park along the VT 140 shoulder at the gravel road and walk a couple hundred feet down to the bridge.

The downstream stretch, accessed by a footpath at the bridge, is very nice. During early spring I use my 8-foot, 4-weight rod, typically with sinking tip, while working my way downstream through pools with small streamers. You can also work back upstream, casting small hair-wing dry flies. Little Black Stonefly (*Taeniopteryx nivalis*) emerge in March along the slow pools. Use a black nymph during that time, such as an A.P. Black Beaver. A variety of mayflies and caddis emerge along the reach later in the season, including good Trico emergences as early as mid-July. I've encountered Hendrickson spinner falls during early May that bring trout to the surface. Sulphur hatches are also encountered a bit later.

LOWER RIVER: WARMWATER SPECIES FISHERY

A Green Mountain Hydroelectric facility operates at the falls but welcomes anglers and recreational canoers. As far back as the late 1800s, power from the falls was used to run mills for 100 years, long before the water was used for hydroelectric power. You can get to the falls and fishing access trail by taking CR 11 off US 4 in Fair Haven to Manchester Road. You'll reach Carver Falls Lane in less than 2 miles. Park at the lot next to the hydroelectric facility and breathtaking falls. An informational kiosk is located at the falls site.

A footpath—wide, very rocky in sections, and very steep—is located 100 feet or so to the left of the electrical facility (you'll see the signage for the path that directs anglers and canoers down into the gorge and the river). It's actually located on the New York side of the river. During higher water flows in early spring, the deep pool at the end of the path can be very difficult for fly anglers to negotiate. Once you latch onto one of many warmwater fish species that inhabit the river below the falls, it could be anything.

Surveys have shown that more than 30 percent of the fish species that swim in Vermont waters inhabit the river downstream from Carver Falls. If historic surveys are considered, the number is around 55 percent. You'll want a 9-foot, 7-weight rod while casting large articulated streamers out into the flow. A 5- to 7-foot sink tip line will get your fly down.

Castleton River

While sampling the Poultney River, or if you were over on the Mettawee, consider a quick jaunt over to its neighboring stream, the 20-mile-long Castleton River, Poultney's largest tributary. The 2002 Vermont Natural History Report, as well as *The Classification of Aquatic Communities of Vermont*, considered Castleton River among the best examples of moderately sized mountain streams in all of Vermont. Castleton River is one of those largely overlooked fly-fishing waters for a couple of reasons. Access is available, but many stretches that are full of trout take some work to get to. I wouldn't call much of the stream prime-looking trout water—some appears to be quite marginal. Good habitat is found along stretches that have brush and overhanging tree limbs, small water that is deceptively productive. I've lost plenty of flies on bank vegetation or negotiating overhead obstructions.

While fishing can be good throughout the seasons, my favorite time to sample its waters is during late summer. In fact, over the past several years, all my Castleton River fishing has been during August and early September. The water is plenty cold enough, and the trout are plenty interested in all sorts of fly patterns. However, as is true on many other similar streams in New York (such as Oriskany Creek), only a single pattern is on my tippet: Chauncy K. Lively's deer hair Carpenter Ant. I've fished that pattern for over 50 years now and have great faith in the fly's ability to excite trout. As I've written in books and magazine articles over the years, my strategy is to cast the ant upstream and purposely create a small wake as it plops onto the water surface.

This brook trout and often sizable wild brown trout stream begins its life in the town of Pittsford, originating on the slopes of Biddie Knob, where it starts flowing south to West Rutland, passing though Whipple Hollow and marshlands. It makes a turn and heads west, roughly paralleling VT 4A on its way, flowing just north of Castleton Village, just

Upper Castleton River during spring, at the North Street Dewey Recreational Fields. Access is easy by parking at the recreational fields that borders the stream. VALERIE VALLA

south of Castleton Corners and Hydeville. From those areas, the stream continues through Fair Haven. Castleton River picks up North Breton Brook along the way and a couple of other small tributaries, such as Pond Hill Brook, before entering the Poultney River. Lake Bomoseen's outlet brook, a principal tributary, makes its short run south to Castleton River in Hydeville.

The best way to access Castleton River is at a few bridge crossings along its path. If you were fishing near the New York stretch on the Mettawee River, head to nearby Fair Haven, located straight up NY/VT 22A, and work your way upstream. If approaching from the Rutland area—fishing Otter Creek, East Creek, or Furnace Brook—start at the headwaters and work your way downstream. In either case, you'll be following the VT 4A corridor to access points.

HEADWATER REACHES DOWNSTREAM TO CASTLETON VILLAGE

Beginning at the headwaters in West Rutland, follow VT 4A west to the Mill Street access point, about 6 miles west of the Whipple Hollow Road/US 7/VT 4A intersection (about 4.5 miles east of Fair Haven). Interesting small water that undoubtedly holds brook trout flows along that distance, but the problem is access. Walking up the railroad tracks that follow the extreme headwaters is too dangerous because the tracks are laid on such narrow and sloped beds.

You're better off first heading to three headwater access areas located in Castleton Village—Mill Street, North Street, and Cemetery Road. Vermont Fish & Wildlife stocks a few hundred brook trout from Mill Street downstream to VT 4A in Castleton Village. Take a right on Mill Street off VT 4A, cross the tracks, and park directly next to an old, small bridge (be careful not to trespass near the surrounding homes). You can't drive across the

By late summer, water levels along the Cemetery reach can lower to a trickle, yet the temperatures remain cold. Trout can often be spotted darting back and forth along this stretch. Stealth and careful approach will result in successful fishing. VALERIE VALLA

bridge, and parking is prohibited in the "Well Zone" area on that side. However, you can walk across the bridge and get into the water. You can also walk along the large field that borders the stream. You'll want a 7-foot, 3-weight rod while casting bead-head nymphs through the riffles. You might pick up a small wild brown trout in the area.

Next, have a look at a couple other access points a very short distance downstream from Mill Street. North Street Dewey Recreational Fields reach, one of my favorite stretches on the Castleton River during spring season, is located just 0.5 mile west of Mill Street. The Dewey fields are located at a bridge crossing off North Street. During spring, access the stream by walking along the left side of the soccer fields to the uppermost area. It can be quite mucky along the stream in early spring, but it's quite nice later on, although by late summer the water levels can drop significantly. However, water temperatures remain quite cold through summer and continue running at good temperatures downstream through the Cemetery Road area.

Cemetery Road reach, located just 0.2 mile downstream from North Street, off VT 4A, is also interesting. Take Cemetery Road off VT 4A to a small bridge crossing that leads to the cemetery. You'll notice a small Pond Hill Brook tributary that parallels Cemetery Road and joins Castleton River at the small bridge. Drive over the bridge and park near a wide trail that follows the stream for a distance. The narrow but often swift-flowing stream continues its journey downstream, directly paralleling railroad tracks. During late summer I like to wet wade the Cemetery reach, casting a 6-foot, 3-weight rod with small dry flies and ant patterns along the riffles and runs. The stream water temperature always remains cold.

While other area trout streams begin to warm by late July, often sending trout into thermal refuge, Castleton River's cold water keeps trout happy. During late summer the Cemetery reach's water levels can drop to a trickle. It's very common to spot trout darting back and forth in the run just upstream from the bridge at that time of year. They're easily spooked; stealth and careful wading are required.

MIDDLE AND LOWER REACHES, HYDEVILLE INTO FAIR HAVEN

By the time the Castleton River departs Castleton Village and arrives in the Hydeville area 2 to 3 miles downstream, it picks up the Lake Bomoseen outlet. Lake Bomoseen itself should be on your list for sure—it's a wonderful waterbody full of fish (see next section). The outlet is only about 0.5 mile long, but it adds more water volume to the now-widening river just downstream from the Blissville Road crossing.

If you don't mind a bit of bushwhacking, make sure you tackle some interesting reaches a few hundred feet upstream from the Blissville Road crossing. Blissville Road intersects VT 4A in Hydeville about 1.5 miles east of the Main Street/VT 4A intersection in Fair Haven. Drive south on Blissville Road 0.3 mile to the bridge crossing. There's a small road shoulder area next to the bridge where you can park at the end of the guardrail. A snowmobile trail joins Blissville Road at the road shoulder next to the bridge (you'll see snowmobile trail directional signage, somewhat out of sight, tacked to a tree facing away from the road). The area is not posted along the snowmobile trail, often overgrown with grass and brush, that leads through a field and provides river access. As a courtesy to the landowner, I always carry a small trash bag and pick up any debris I might encounter in such areas. The stream has slow-moving pools, but a bit of woody stream structure can be encountered upstream. Brook trout and a few browns inhabit the stream reach.

From the Blissville Bridge crossing, the Castleton River flows just another 5 miles or so before it joins the Poultney River. Along its final mileage, it enters Fair Haven Village. If you're after hatchery brown trout, head downstream into Fair Haven. VF&WD typically

Brook trout inhabit upper Castleton River's cold headwaters. They'll hit small dry flies and terrestrial patterns during late summer.

Carpenter Ant patterns work well on the Castleton River during late season.

stocks a few hundred browns in the section downstream from the VT 22A crossing in Fair Haven. At the VT 22A (Main Street) bridge crossing in Fair Haven, park downstream from the bridge on what was once the Adams Road bridge crossing.

The former bridge is blocked off with concrete barriers, but this is a good place to view the dam and the cascading falls below it. A footpath leads to the river at the far side of the former Adams Street bridge, but make sure you're not walking through private property. The Castleton River courses downstream through runs along a ravine area that holds fish. One drawback on this reach is the existence of massive amounts of unsightly slate slabs scattered along the sloped riverbank. Use a slightly longer rod, I use an 8-foot, 4-weight, swinging small streamers through the runs. The water can be quite high during early season, and very difficult to negotiate and fish.

Lake Bomoseen

While fishing the areas trout streams, Poultney and Castleton Rivers, consider a warmwater species outing on two of the area's popular lakes. The largest lake that lies totally within Vermont's state boundaries, 2,400-acre Lake Bomoseen provides a variety of warmwater fly-fishing opportunities. The lake's short outlet feeds the Castleton River near Blissville Road. Situated in the towns of Castleton and Hubbardton, public access is located at Bomoseen State Park on the lake's western shore and at formal VF&WD boat launches at the north and south ends of the lake. Bomoseen State Park is located at 22 Cedar Mountain Road in the Fair Haven area. Camping is available at one of the 55 tent/RV sites at the park. The southeast corner of the park, off the point, is a popular fishing section that harbors a variety of gamefish.

VF&WD biologists consider Lake Bomoseen one of the best fishing lakes in the state—given its mixture of bass and trout populations. Sizable smallmouth and largemouth bass ply its waters. Fisheries biologist Shawn Good can't say enough good things about Bomoseen's fantastic angling opportunities, including for those casting flies. Shawn provided some great insight:

"As a fisheries biologist with VF&WD for 27 years now, I can say that Bomoseen is quite possibly one of the most productive fisheries in the state for quality, for a number of species. Strictly speaking about bass, I don't think there's any other lake in the state that produces more or bigger largemouth than Bomoseen. And the smallmouth fishing is fantastic too. I've seen more 6- to 8-pound largemouth, with some exceeding 9 pounds, here than anywhere else in my surveys."

Lake Bomoseen in the Castleton area is regarded as one of Vermont's best bass fisheries. Trout also inhabit the lake.

Shawn obtains data on bass populations using electrofishing techniques at night. He encountered a surprise a couple of years ago: an American eel that turned up with bass. Some might wonder how a few American eels, one 4 feet long, turned up in an inland lake. Shawn Good explained exactly how that happens. The eels spawn in the Sargasso Sea near Bermuda, migrating up the eastern seaboard to the St. Lawrence River to Lake Champlain to the Poultney River to the Castleton River to Lake Bomoseen outlet and finally into the lake, where they grow to adult size. They spend 4 or 5 years in the lake then migrate back to the Sargasso Sea, where they spawn then die. It's unlikely, but should you come across a 4-foot-long baseball bat–girthed creature while bass fishing, it's not a python.

Bass-fishing anglers were among the first to rally against a plan to use herbicides to control Eurasian water milfoil, an invasive plant not particularly welcomed on the lake by recreational users other than anglers. While obnoxious to a point, the milfoil provides cover for forage baitfish. Local surrounding towns also helped squash the plan. The Vermont Department of Environmental Conservation denied an application by a lake association to apply herbicides.

While Bomoseen Lake's bass population is the likely target species for both fly rodders and conventional tackle anglers, deeper areas of the lake around Neshobe Island support trout. VF&WD stocks some 3,000 browns annually in the lake. After populations fell off, smelt populations have regained a presence due to stockings by the state. Smelt provide a good forage base for the up to 10-pound browns that grow in the lake.

Rowboats, kayaks, and canoes can be rented at Bomoseen State Park. If you're hauling in your own watercraft, you'll have easy access at the Edward F. Kehoe launch off Creek Road on the south end of the lake. Located off Creek Road, 1.9 miles from the US 4A/Creek Road intersection in Castleton, the Kehoe launch area is very busy with large fishing boats and other recreational congestion during summer. The Thomas Evanoika Access launch on the north side of the lake, located at the south end of Johnson Spooner Road, is my preferred access point while fishing out of a canoe. If you're targeting warmwater bass and panfish, you'll run into fewer of the large boats that frequent other areas of the lake. From the launch access, paddle or row beneath Floating Bridge Road to a lily pad area. Shawn chimed in on this:

"Because of the shallowness and the vegetation, it doesn't get the boating traffic the main lake does. It can get pretty thick in the summer, but there is a 'river channel' that runs from north to south from the inlet of the stream that feeds the north wetland and the lake. You can see it quite well on satellite imagery on Google Earth. Fish use this channel to move throughout the wetland above Float Bridge. Bass fishing is incredible up here, particularly right after ice-out. In fact, as soon as the ice goes out, there's no better place to fish on the lake than above Float Bridge, because the shallow depth and dark detritus and organic matter cause the water in the wetland to warm up earlier than the rest of the lake. So you can find big schools of active fish moving into the wetland to warm up their body temps after a long cold winter. That's where feeding activity starts first. Whether it's large- and smallmouth bass, black crappie, yellow perch, pumpkinseed and bluegill, northern pike, bullhead—you can catch them all up there in April, even late March, depending on when ice-out happens. As for techniques, anything goes. If you're all about fly fishing, there's no reason that won't work."

Shawn was particularly intrigued about one warmwater fish species that inhabits lake Bomoseen's north end—Redfin pickerel. And in his opinion, fly-rod fishing could very well be the best tactic to get into one of those fish:

Bomoseen's north end, at the Floating Bridge Road crossing, is much better suited to fly rodders than the more expansive areas around the lake, especially on the south end. Good habitat can be found, including lily pads, which provide ample cover for bass, pickerel, and other warmwater species.

"One unique thing about Bomoseen, and the north wetland above Float Bridge in particular, is that it's ground zero (as I call it) for one of the rarer species of the Esocidae family in Vermont. Bomoseen has THE BEST and most robust population of Redfin Pickerel anywhere in Vermont, certainly and quite possibly anywhere in the Northeast. These little members of the pike family max out at 10 or 11 inches, but it's a Life List species for many. They live in small pockets of open water in the thickest, densest beds of aquatic vegetation they can find, and 99 percent of the population in Bomoseen is above Float Bridge. It seems to me that a determined fly angler would have the best luck catching these compared to other techniques. Most anglers never see them, or don't even know they exist. The rare angler that catches one and emails me a photo asking, 'What did I just catch?' is usually ice fishing in the wetland with small (1/32 or 1/64 ounce) jigs and microplastics for panfish. Just the right sized fly, fished in pockets around the heavy vegetation in the evening, I would image would have a very good chance of catching one. I see them by the hundreds when I'm doing fisheries surveys up there."

Orange Peel Optic Bass and Cherry Peel Optic Bass *Tied by Mike Valla*

These two colorful bass flies have performed well for the author over many years, fished on a variety of warmwater ponds and lakes. The flies sink fast with heavy Partridge 2/0 Predator hooks. Cashmere goat is used on these two flies, but they can also be tied with marabou wings or any material that provides motion. Syracuse Orange Lion Brand yarn (#133) and Cincinnati Red Lion Brand yarn (#113) form the bodies. They are both ribbed with large oval gold tinsel. The eyes are created with Testors paints.

Lake St. Catherine

Lake St. Catherine, a short drive from Lake Bomoseen, is also a popular bass destination. Composed of a chain of three separate waterbodies, 852-acre Lake St. Catherine is one of the most popular multiuse recreation lakes in southern Vermont. The smallest of the linked waterbodies, Lily Pond, is at the north end. It drains directly into Lake St. Catherine, which in turn spills out into Little Lake. Little Lake spills over a small dam, continuing its flow as Mill Brook Creek, which eventually joins the Mettawee River.

Lake St. Catherine's proximity to the Mettawee, Poultney, and Castleton Rivers provides an additional nearby destination for fly anglers who are visiting those streams. You can reach the lake by driving to 3034 VT 30 in Poultney. Public boat, kayak, and canoe launches are available at 117-acre Lake St. Catherine State Park. Another launch at the VF&WD boating access location is available at the bridge that separates Little Lake from the main waterbody.

Lake St. Catherine is a perfect destination for anglers who wish to include other family members on a swimming, boating, and fishing outing. Those interested in camping can reserve one of the 50 tent/RV sites or one of the 11 lean-to sites. Anglers need not be concerned about dragging their own watercraft to the lake. Rowboats, kayaks, and canoes can be rented at the park. I introduced my family to the park nearly 30 years ago; it's still a favorite among many of our friends.

Brown and rainbow trout, largemouth and smallmouth bass, pike, panfish, and other species swim in its waters. Rainbow smelt, golden shiners, and emerald shiners provide a good forage base for salmonids and warmwater species. The lake has a maximum depth of 68 feet. Head to the south end or north end of the lake if you're targeting bass, northern pike, or panfish. Deep water along the west side of the lake provides good habitat for trout. VF&WD typically stocks 3,000 brown trout and around 1,000 rainbow trout annually during April.

Big Branch Stream

Big Branch Stream, a tumbling boulder-strewn, crystal-clear wild trout stream, is home to brookies as well as brown trout. In mid-December 2023 I performed a DNA sample on the stretch upstream from Silver Bridge (about a mile upstream from its confluence with Otter Creek). Results showed a significant amount of brown trout DNA coming through that reach, with 100 percent base-pair matching. Late summer and early fall, during low water, are prime time for 3- to 4-weight rods and a variety of effective flies, including small flashy, colorful streamers. Bead-head patterns also work well.

The little stream courses through a gorge-like valley flanked by steep mountain slopes. It flows through a mature northern hardwood forest that provides habitat for black bears, white-tailed deer, moose, and other wildlife. Flowing westerly out of the mountains, Big Branch Stream eventually dumps into low-gradient Otter Creek, another popular Vermont fishery (see page 62). In addition to Big Branch, the wilderness area offers anglers ample backpacking opportunities to explore other connecting brookie-filled waters, such as Little Rock Pond and Griffith Lake.

A tributary that feeds Otter Creek's headwaters, Big Branch Stream arises in the serene White Rocks National Recreation Area and the Big Branch Wilderness in Vermont's Green Mountain National Forest.

The ponds, as well as the more isolated Big Branch stretches, can be reached via a section of the Appalachian/Long Trail that crosses the region. Big Branch's lower sections are accessible from FR 10 via US 7, the major road that runs the length of eastern Vermont. If you are connecting to US 7 from the south, stop by the Orvis flagship store in Manchester. You'll pass it on your right. Pick up VT Historic Route 7A out of Manchester Center (a bypass) and drive another 5 miles to merge onto US 7

Colorful brook trout can be encountered on Big Branch Stream from early September into October.

North toward Rutland. Drive close to 8 miles and turn right onto Brooklyn Road in the hamlet of Danby.

Brooklyn Road, which continues as FR 10, heads up the mountain to Big Branch Stream, passing over Otter Creek along the way. A turnout at the small Otter Creek bridge crossing marks the upper limit of Otter's Special Regulations section (see Otter Creek, page 62). The state stocks "trophy-size" 2-plus-year-old browns and rainbows in the Special Regulations area.

Big Branch Stream will appear on your left as the road begins its climb. You'll encounter a nice turnout parking area at about a mile past the Otter Creek crossing at what's called the "Silver Bridge." An informational kiosk at the turnout introduces visitors to the Big Branch Stream area. A path from the parking area leads down to the creek and is the easiest access to Big Branch Stream. If you want to sample stretches well upstream from this point, you'll be in for a good hike—or an overnight camping adventure.

Once you drive up the winding road past the Silver Bridge, the stream drops out of sight into a deep gorge. In a little over 2 miles past the Silver Bridge, you'll find a paved parking lot and two small bridge crossings over small brooks, just a couple hundred feet apart. The first small brook, just before you reach the parking lot, is Little Black Branch. The second small brook—Big Black Branch—runs along the parking lot. Both brooks, Big Branch Stream tributaries, support small wild brook trout and sometimes small brown trout. The parking lot is an Appalachian Trail cross point. A footpath at the far end of the parking lot leads directly into Big Black Branch if you're interested in fishing through stretches downstream from the bridge crossing.

Incidentally, you'll see an informational kiosk across the road from the parking area, complete with an area map showing the various trails to popular hiking areas. The trailhead to 18-acre, spring-fed Little Rock Pond begins at the kiosk. The moderate 2.2-mile trek begins on the banks of Big Branch Brook but quickly veers off toward Little Black

Branch and follows it to the brook trout pond that's stocked annually with 2,000 fingerlings. I haven't experienced it myself, but I'm told the pond has a *Hexagenia limbata* hatch during late June.

Big Black Branch tributary, located at an Appalachian Trail crossing, is inhabited by small wild brook trout, along with a few small browns. It can be accessed at a parking lot off FR 10, about 2 miles upstream from Silver Bridge. Big Black Branch's sister brook, Little Black Branch, is located just before the parking lot. VALERIE VALLA

Getting back to Big Branch Stream itself, and a prospective trek into remote upstream reaches, you have two choices: a day trip or an overnight backpack excursion. An overnight camping trip would be great, but a day trip is fine too. In either case, you'll reach Big Branch Stream's upper reaches by taking the Appalachian Trail. Park your vehicle at the lot, then walk up FR 10 a couple hundred feet to the trailheads. National Geographic Green Mountain National Forest #748 is very useful. The map clearly shows important details in the area, including trails.

The first trailhead, marked by "AT" signage, directs hikers who are heading north. You want the second trailhead, located less than 100 feet away, also marked by AT signage that directs hikers to Griffith Lake. Count on 40 minutes of leisurely trekking to reach the stream. If asked the best time of year to sample the upper reaches via the AT, I would say summer into early September. For one thing, the water levels are usually lower that time of year, making it a bit easier to fish the small pools by boulder hopping. The other advantage is that the stream can be wet-waded; you can leave the heavy waders behind. Years ago I used to trek more than a few miles into the backcountry with waders on! Those days for me are forever gone!

Several hundred feet down the trail, you'll encounter a "sign-in" kiosk. For safety reasons, it's important to sign in and indicate where you're heading. Green Mountain National Forest staff routinely check the sign-in book. The trail begins as a mostly dirt path with a few rocks here and there, but it soon goes through stretches that are rocky and real ankle-sprain traps for sure! Some rough sections have been taken care of with plank board crossings. A wading staff is useful. Springtime freshets that race down the mountain from above create all the washed-out areas strewn with rocks. Through the entire trek, Big Branch Stream is out of sight.

A suspension bridge along the Appalachian Trail spans Big Branch Stream several miles upstream from Silver Bridge.

In less than an hour of trekking through the forest canopy, you'll arrive at a lean-to shelter, where AT trekkers or backpackers often spend the night, and a suspension bridge that spans Big Branch Stream. Lean-to shelters are open to public use—no reservations or fees are required. While it's a first-come, first-served basis, etiquette requires offering space to tired AT trekkers who might come along and need a place to rest their bodies and souls. The shelter, open on the front side, has room for about eight persons. Once you get to the lean-to, you'll be able to get into the water without too much difficulty. The striking suspension bridge that spans the stream, the site of an old lumber mill, is a short walk from the shelter. You'll see remnants of the old operation.

I actually prefer the upper reaches of Big Branch Stream because the gradient along the small pools just downstream from the suspension bridge is a little gentler than down at Silver Bridge. But again, that's why I fish the reach during lower water levels in late summer. Be aware of the numerous rocks and boulders you'll be crawling around while dipping your favorite brook trout fly pattern through the small pools. You don't want to need to be rescued because of a sprained ankle or serious injury.

Otter Creek

Roughly 112 miles long, the longest of Vermont's rivers contained totally within the boundary of the state, Otter Creek is truly a stream of varied moods and promises when it comes to sport fishing. The stream begins in the Green Mountain National Forest as a small water then flows into Emerald Lake, not far from Manchester. From Emerald Lake, Otter Creek flows northeasterly as a slow meandering stream, passing small towns such as Mount Tabor, Danby, and Wallingford, as well as larger settlements and cities. It courses over hydroelectric dams along its sinuous path.

Middlebury Falls is one of many falls that are strung along Otter Creek's 112 miles.

In addition to the previously discussed Big Branch Stream (see pages 59), which flows out of the Green Mountain National Forest and joins Otter Creek in the Mount Tabor–Danby headwater area, Otter Creek gathers a couple of other small trout tributaries as it flows through the Rutland and Pittsford areas. East Creek Furnace Brook and Neshobe River are frequently fished by fly anglers. Larger trout waters such as the New Haven and Middlebury Rivers wed Otter Creek farther downstream.

Many miles downstream from its source, Otter Creek flows into the town of Middlebury, where the stream, now of greater volume, crashes over historic Middlebury Falls then pushes onward to the city of Vergennes. On its way to Vergennes, the stream passes through one of the most picturesque areas along its course. Beldens Falls, located downstream from Middlebury, flows through a beautiful natural area that's popular with hikers and day users; the falls are accessed at the Beldens Hydroelectric Station located at 402 Belden Road in New Haven. New Haven River, a trout fishery, joins Otter Creek a short distance downstream from Belden Falls.

In Vergennes, Otter Creek crashes again over a dam, located not far from its terminus at Fort Cassin Point on Lake Champlain. The lowermost sections downstream from Vergennes, closer to its confluence with Lake Champlain, are known by anglers as prime habitat for a variety of warmwater species. Everything from pike to bass to carp and other scrappy warmwater fish swim in its waters.

In many ways, fly-fishing for warmwater species, including pike, surpasses trout fishing experiences available from Middlebury upstream to its source. Brian "Lug" Cadoret from Cornwall, Vermont, co-owner of Stream and Brook Fly Fishing, is an enthusiastic Otter Creek pike angler. "May into early June is prime time for pike on the Otter Creek," Brian said during a short outing we shared on the creek. However, trout fishing is fully enjoyed by fly fishers on Otter Creek who prefer to latch into rainbow, brown, and a few brook trout

Otter Creek reach at Florence Avenue bridge crossing in Wallingford. Upper stretches like those in Wallingford have much more favorable trout structure than most other stretches, both upstream and downstream.

that are stocked in its headwaters. Of course there's no telling where on Otter Creek you might unexpectedly latch into a trout.

A 2-mile-long Special Regulations Trophy Trout section can be accessed near the Mount Tabor–Danby town line, about 13 miles from Manchester Center if you're coming from that area, 6 miles north of Emerald Lake State Park on US 7, and 1 mile north of the Brooklyn Road/US 7 intersection, the Trophy Trout section's upper boundary. You can also get into the stream at a small bridge crossing back on Brooklyn Road. Otter Creek Wildlife Management Area (WMA) signage marking the parking spot is on US 7. Much of Otter Creek's appearance along its Trophy Section mileage appears dull, slow moving, and uninviting, although a few faster runs are available. I've fished Otter Creek from the WMA parking area all the way upstream to the confluence with Big Branch Stream a few times (see page 59) and haven't been overly impressed. Since the Trophy Section is stocked annually with large 14- to 17-inch brown trout and rainbow trout, the reach is heavily visited by bait casters during May, when the hatchery trucks pull in to deliver their loads.

Leaving the Mount Tabor–Danby WMA area behind, I much prefer sections downstream that flow through nice rocky runs and riffles in the Wallingford area. The Hartsboro Road reach at the Wallingford town line is very nice and is popular with fly fishers. That section receives stockings of 11-inch rainbows. It's located a quick 4 miles north on US 7. Turn right on Hartsboro Road just beyond the stone quarry. You can park at the far side of the bridge crossing, near the railroad tracks. This is good water to dabble Stonefly nymphs and Hare's Ear nymphs. Elk Hair Caddis dry flies can elicit strikes. May and June are good times to fish that reach, again in September, although I've also trekked along the stream section during nice winter days when I can have it all to myself.

Another nice stretch is located just up the road off VT 140. If approaching from the south, turn left at the VT 140/US 7 intersection in Wallingford. Follow VT 140 for a short distance to a bridge crossing. Park at the Wallingford Recreational Park near the bridge, and fish small streamers through the runs downstream to another bridge crossing on Elm Street. By the way, if you're on the VT 140 bridge reach and fishing is slow, consider taking a 10-mile drive along VT 140 to the small town of Middletown Springs, where you can try the Poultney River, a brook and brown trout stream (see page 49). Brook trout are stocked along the headwaters near Middletown Springs.

Fly anglers who don't seem to mind casting flies in an unattractive congested area in center Rutland head to Meade's Falls. VF&WD annually stocks a few thousand brown trout from center Rutland to Clarendon and center Rutland to Proctor. The falls are named for James Meade, Rutland's first settler, who came to the falls in 1769. Settling in a log cabin, Mead built saw- and gristmills at the falls. Check out the historic marker at the falls that's quite easy to access.

If approaching from Rutland, you can get to Meade's Falls by turning onto US 4 at its intersection with US 7. In West Rutland take exit 6, then turn onto US 4 Business East and continue to a right on Simons Avenue just before crossing Otter Creek. You can park at the far end of Simons Avenue just before it continues to the right. A footpath leads to the stream from the road. It's unfortunate that individuals have trashed the area with litter, something that is perhaps inescapable along a populated area. The best times to target this area are late May into early June and again during the fall. A 9-foot, 5-weight rod is a good choice for this reach where you'll want to cast streamers.

Otter Creek's character changes downstream from Rutland. Not all that far from Meade's Falls is Proctor Falls, about 5 miles downstream. From Meade's Falls, cross the bridge and take a quick left onto VT 3N. Take VT 3N to Patch Street, which leads to a nice area directly

Otter Creek near Brandon meanders slowly through areas that provide good pike habitat and good fishing during early season. VALERIE VALLA

next to the river and its falls. It's a nice place to fish primarily for pike and smallmouth bass, but trout also inhabit the area. Picnic tables are at the site, convenient for a lunch break.

Otter Creek's mileage downstream from Proctor, in the Brandon area, can be characterized as a very slow-moving river that wanders through areas removed from traffic and congestion. Sections tucked away off paved roads provide good pike habitat and are productive during early spring. Pike and bass fishing can be quite good for many more miles downstream. "The pike and smallmouth bass are good from Proctor Falls all the way to below Vergennes Falls, which is probably 45–50 miles of river!" Brain Cadoret noted.

When fishing Meade's Falls, Proctor Falls, or reaches near Brandon, consider sampling three popular tributaries that are close by: East Creek, Neshobe River, and Furnace Brook.

East Creek

A 5-minute drive from Meade's Falls will get you to East Creek, an Otter Creek tributary popular with trout anglers. You'll encounter big stocked triploid rainbow trout by mid-May. East Creek's Special Regulations Trophy Trout section runs 2.7 miles from Patch Dam downstream to the confluence with Otter Creek. You'll first want to sample East Creek along its lowermost reaches not all that far upstream from its confluence with Otter Creek by driving to Pine Hill Park, located at 2 Oak Street Extension in Rutland. Park at the Giorgetti Athletic Complex. The Pine Hill Park reach flows through a wooded area of nice riffles. Nymphs work well for enticing the rainbows.

If you fish downstream from the wooded Pine Hill Park area as far as the Crescent Street bridge crossing, which isn't that far, you might have a look at the stream along the Rutland Creek Bike Path. It's at the end of Earl Street, which is off Crescent near the bridge crossing. Parking isn't available at the end of Earl Street, so either park upstream and walk to the

East Creek at the North Grove Street bridge crossing, a short distance downstream from Patch Dam. This reach flows through a Special Regulations section. It's well stocked with rainbows.

trailhead or, better yet, fish downstream from the Crescent Street bridge crossing and pick up the bankside trail on your way down through the riffles and runs.

Another reach a short distance upstream is easily accessed at the North Grove Street bridge crossing. It's located just beyond Rutland Country Club, which is also off North Grove Street. Parking is available at the bridge crossing, but make sure you don't block the gates to the Patch Dam facility. If you want to fish the pool below the dam, walk around the gate, if it's closed, and along a gravel road that ends at the dam. VF&WD has posted signs along footpath that leads down a hill to the stream to remind anglers of the Special Regulations. Unlike the majority of Vermont trout streams, this reach is not open to year-round fishing. A headwater section that runs from Holden Road to Sangamon Road in the town of Chittenden receives brook trout stockings.

Furnace Brook

If I had a choice of fishing Otter Creek trout tributaries near and not far from the Rutland area, selecting Furnace Brook would be a no-brainer. A 15-minute drive from East Creek crossing on north Grove Street will deliver you to Furnace Brook's upper section in Pittsford. If you're at East Creek, take north Grove Street to US 7 to Pittsford. Furnace Brook is a small unstocked trout stream that's born as a tumbling brook in the Green Mountain National Forest north of Pittsford.

Furnace Brook begins after gathering a few small brooks that combine in the mountains as it makes its way south along River Road. The brook continues its journey along Dwight D. Eisenhower National Fish Hatchery lands in North Chittenden. Atlantic salmon, lake trout, and brook trout are raised in the hatchery that's been in existence since 1909. Atlantic salmon have been raised in the hatchery since 1980 as part of the Lake Champlain

Furnace Brook, a few miles upstream from the Dwight D. Eisenhower National Fish Hatchery. The stream supports wild brook trout, brown trout, and rainbow trout. VALERIE VALLA

Restoration Program. The hatchery, located on Furnace/Holden Road, welcomes visitors year-round and is easy to find. It's located just about 4 miles from the VT 3/US 7 intersection in Pittsford Mills. Take Furnace Road (about 0.5 mile from the intersection). In just over a mile, make a sharp right on Furnace Road where it connects with Hitchcock Road. In about 2 miles you'll see the hatchery entrance on the right.

A bit of folklore surrounds the time when then-president Dwight D. Eisenhower fished Furnace Brook without landing a trout. The story that now-retired Vermont journalist Bob Bennett reported on a couple of years ago in the *Rutland Herald* newspaper was how Ike was scooped by a local kid during his June 1955 visit to the area and stream. The young lad heard that some of the large brook and brown trout held at the then-named Holden Federal Fish Hatchery would be pre-planted in a Furnace Brook hole, without Eisenhower's knowledge, to provide a little leverage for the president's fishing success. The kid snuck in there prior to Ike's arrival with his new cane rod, a gift from Orvis. The kid caught all the fish using a cheap Montague rod and slipped away with the batch in his rucksack.

The small stream departs hatchery acreage then continues its path past Pittsford, where it flows under US 7 near its intersection with VT 3. Pittsford Mills itself has a place in fly-fishing history. An old yet effective dry fly named Spirit of Pittsford Mills was featured by Catskill fly tier Elsie Darbee and is still promoted today by Murray's Fly Shop in Virginia.

Continuing south, the stream widens after accepting inflow from Sugar Hollow Brook (also an unstocked trout stream). A short distance later, Furnace Brook passes under Cooley Covered Bridge on Elm Street then, in a short time, enters Otter Creek. Cadwell Trail, a public nature trail that intermittently follows Furnace Brook and surrounding meadows, is accessed at Cooley Covered Bridge. Signage and parking are on Elm Street, next to the bridge. Just upstream from Cooley Covered Bridge, the stream flows under Gorham Covered Bridge, also located off Elm Street. During spring you'll often see conventional tackle anglers at the Gorham Covered Bridge crossing. Furnace Brook during summer is moderately wide yet slow moving, flowing over an unappealing silty bottom as it makes its final run before entering Otter Creek.

Lower Furnace Brook reaches are not my favorites, but sizable brown and rainbow trout make their way up into the stream from Otter Creek and then actually quite far upstream. I recently had a nice chat with Shane Hanlon, hatchery manager/project leader at Dwight D. Eisenhower National Fish Hatchery. Shane and a couple of his staffers explained that while Furnace Brook isn't stocked, the stream supports brown, rainbow, and brook trout. From about the hatchery upstream to its sources, you'll typically run into only natural-spawn brook trout. I performed a DNA sample just upstream from the hatchery, and data indicated the presence of all three species: brook, brown, and rainbow trout.

A good place to get into your waders is located just before you reach the hatchery entrance. There's a wide shoulder area that can accommodate a couple of vehicles directly at streamside. More remote areas that tumble through beautiful forest canopy settings are located upstream. From the hatchery entrance, drive 0.1 mile to River Road, on the left. River Road is a gravel road that's easily negotiated with a regular vehicle, although I usually take my Jeep Wrangler up the road when driving to its extreme headwaters where River Road meets Green Mountain National Forest lands. Small dwellings are scattered along River Road, which follows Furnace Brook, crossing the stream here and there along the way, but they become more separated as you continue driving.

Spirit of Pittsford Mills *Tied by Mike Valla*

A classic dry fly created by Stephen Belcher II, named by his daughter Jane, was enticing trout on Furnace Brook. The pattern was tied with duck-down body, palmered clipped ginger hackle, and grizzly hackle point wings. It was a favorite of Catskill fly fishers back in time. Harry Murray of Murray's Fly Shop in Edinburg, Virginia, called it an early-season go-to dry fly.

About 1.6 miles up River Road from its intersection with Holden Road near the hatchery entrance, a nice wide parking shoulder is located at a double bridge crossing. It provides a good access point and a nice place to get into the water. Farther upstream you'll run into posted green signs that prohibit hunting and trapping activity next to yellow signs that are clearly altered to make it known that fishing is allowed. Fish small nymphs through the pocketwater, and cast small elk hair dry flies to entice wild yet small brook trout.

Neshobe River

A tributary of Otter Creek, Neshobe River's journey begins from springheads located high in Green Mountain National Forest at the border of Windsor and Addison Counties near Mount Horrid. From its more than 3,000-foot elevation, the young stream flows briefly southerly in its Goshen headwaters, then turns west once it hits VT 73 (Gap Road). Neshobe River is still small water as it roughly follows VT 73 until its picks up North Branch of Neshobe River as it crosses into Rutland County. A short distance downstream from the Churchill Road bridge crossing, Leicester Hollow Brook in the Forest Dale area also adds to the stream's volume.

Neshobe River upstream from the Churchill Road crossing reach tumbles through nicely forested, well covered area that provides good habitat among its cobble and small boulders.

Neshobe River, still a brook or small creek size, pushes for a few miles toward the artsy village of Brandon, passing through Forest Dale, changing its character along the way. Along its main stem Goshen headwaters, the stream is high gradient and confined, its cobble and small boulders mixed with small pools and runs. Once the river enters reaches in Forest Dale, downstream from Churchill Road crossing, its steep slope moderates as it enters agricultural fields along the Town Farm Road crossing. Its character changes again once the stream hits Brandon, where the Neshobe River tumbles over a small laid-stone dam in the village then crashes over the main falls a short distance away. Downstream from the falls, Neshobe River loses its riffles and run structures and quiets, flowing as a gentle sandy-bottomed stream through the low-gradient Champlain Valley before it finally joins Otter Creek.

Access is largely available at one of the dozen bridge crossings along the stream. Some crossings provide better fly-fishing access than others. Uppermost headwaters along Gap Road in Goshen (VT 73) are way too small to effectively fish flies. I once ventured upstream as far as the South Hill Road crossing in Goshen, only to be greeted by flow that was quite small for effective fly rodding. You're better off getting into a couple favorite bridge crossing access points downstream from the Neshobe River's confluence with North Branch of Neshobe River. During certain times of the year, you'll also want to sample the water downstream from the falls in Brandon.

VF&WD regularly stocks the Neshobe River with a token number of brook trout, around 500, from the village of Brandon to Forest Dale. In correspondence, VF&WD fisheries biologist Shawn Good said, "The Neshobe has great brown trout and brook trout fishing (especially the North Branch east of Brandon)." However, you could run into one of many fish species that at one time or another inhabit Neshobe's mileage, depending on what reach you're fishing, from its source to its confluence at Otter Creek.

According to Good, the main stem of Neshobe River has 18 historical/monitoring sampling sites. While it's been a few years since VF&WD surveyed the Neshobe River and North Branch of Neshobe River, trout population data taken over the years is worth mentioning. In 2012, at the 355-foot-elevation station, a good number of mixed age-class brown trout were noted, with very few brook and rainbow trout. At the 566-foot elevation, brook trout predominated in higher numbers, along with higher numbers of rainbow trout estimates per mile than found at the lower-elevation site. At the 700-foot-elevation sample site, all three trout species were found—although very few browns in the mix estimated per mile. At the 900-foot-elevation site, brook trout again predominated the population estimates per mile. Mixed-age classes of rainbows inhabited the reach, but very few browns. Not surprisingly, brook trout held strong at the 1,240-foot elevation, with a few rainbows.

Not having any updated VF&WD data, I wondered if high rain and flooding events might have changed trout population dynamics. In September 2024 I took DNA samples on both the main stem at Town Farm Road and on the North Branch just upstream from the VT 73/Town Hill Road intersection. Sequencing revealed an abundance of 100 percent base-pair matches for brook and rainbow trout DNA flowing downstream through the North Branch site, but no brown trout DNA was detected. Much lower values of rainbow and brook trout DNA flowed through at the Town Farm Road crossing, although a small amount of brown trout DNA was detected.

I was somewhat surprised that rainbow trout DNA showed up in my North Branch sampling. For a second I thought maybe rainbows had come that far upstream from Otter Creek. However, given there are two impassable falls in Brandon, that reason didn't seem likely. Even during the high rain events and substantial floods that have occurred over the years, it would be unlikely that even the most ambitious rainbow could get upstream. I forwarded the North Branch DNA sequencing results to VF&WD fisheries biologist Shawn Good, and asked how the heck rainbows got in there to begin with.

"They probably originated from stocking in the heyday of 'spread all fish species everywhere and see what happens' attitude most fish and game agencies worked under in the early to mid-1900s," Shawn theorized.

NORTH BRANCH TO CHURCHILL ROAD CROSSING

If you're interested in trying the more tumbling upstream stretches with plenty of pocketwater and small pools, head to the North Branch of Neshobe River and main stem Neshobe River stretches that flow down through a secluded, confined area to Churchill Road crossing after the two streams combine. Access the North Branch by driving about 5 miles north on VT 73 from the VT 73/US 7 intersection in Brandon to a left turn on Town Hill Road (FR 32). The small water flows along Town Hill Road for a distance from the small bridge crossing next to Goshen Town Hall. Parking can be tough but possible along a couple of road shoulder areas.

After a few casts on the North Branch, you can continue downstream from its confluence with the main stem of the Neshobe River by getting into the water at Churchill Road crossing. It's located off VT 73, less than a mile downstream from the VT 73/Town Hill Road intersection (or just short of 4 miles upstream from the VT 73/US 7 intersection in Brandon if approaching from that direction). The narrow, bumpy dirt road that passes one dwelling is marked "Dead End" but is passable along the short distance to the small bridge that has undergone repairs after floods ravaged the area. You'll see a large "No Public Parking" sign as you approach the bridge; park on the far side of the bridge, making sure not to block the road. Work the water upstream, with most any flies you'd select while fishing similar step-pool waters.

TOWN FARM ROAD CROSSING

After fishing the tumbling reaches upstream, head downstream to moderate gradient reaches along the Town Farm Road bridge crossing in the Forest Dale area. (On your way downstream to Town Farm Road, you'll pass VT 73 intersections with Furnace Road and North Street. Bridge crossings are located on those roads, but access isn't as easy.) Town Farm Road is located off VT 73 (Forest Dale Road), a couple of miles south of Churchill Road (about 1.5 miles north of Brandon if approaching from that direction). Drive past the golf course to the bridge crossing.

Access directly into the water and its downstream runs is quite easy at the bridge. The last time Valerie and I explored that reach, we were greeted by the landowner who maintains the small grassy area next to the bridge and stream. Instead of having us park on the busy shoulder across the road, the landowner directed us to park on her lovely mowed property at streamside—a much safer area. Try the pools and runs a hundred feet or so downstream from the bridge. The Neshobe River flows through an area downstream from Town Farm Road, crosses the Neshobe Golf Club course, then continues past Stone Mill Dam Road crossing, where access is difficult. Access is better at the next crossing at Wheeler Road, 0.5 mile downstream.

BRANDON WATERFALL

After the stream passes Wheeler Road, it enters residential areas before hitting the first, small falls in the village of Brandon at Kennedy Park, a couple of miles downstream. The Neshobe River then passes beneath Center Street (US 7), out of sight, then emerges at Green Park almost directly across Center Street from Kennedy Park. Green Park has a pavilion with a picnic table, along with a viewing area of the striking large waterfall that tumbles down between buildings.

Access for fishing the large plunge pool beneath the large falls is available largely by courtesy of property owner Jeff Haylon. Jeff, one of the nicest guys you'll ever run into, doesn't own the falls, but his property just around the corner from Green Park overlook provides a quick way into the water. A quick walk down Briggs Lane (it's next to a quaint bookstore named The Bookstore) will get you to Jeff's labor of love. You'll spot a large board-and-batten structure Jeff renovated and is quite proud of.

"The shop was built sometime pre–Civil War, likely as part of Roger Fuller's manufacturing operation and distillery sometime in the 1820s–1840s," Jeff explained. "It became the on-site farrier and repair shop. It fell into disrepair sometime after the Civil War until it was bought by John Richmond, who was a full-time farrier until his death in the 1950s. From then it was disused by a series of owners until I bought it in 2021 and fixed it up to what it is now. The unconfirmed local lore is that most of the horses that went south from Brandon and north in Vermont got shoed at the shop on their way to fight in the Civil War. I choose to believe it."

Anyway, please park up on Center Street somewhere and walk down Briggs Lane to Jeff's workshop. Go to the far side of the building and look for a small footbridge that leads to the water. Along a footpath that leads to the water, you'll encounter private property signage on the left side on an abutting neighbor's property. Just continue along the footpath to the Neshobe. Once you're in the water you're fine. I haven't fished the pool and runs downstream from the falls myself, but I've been told the fishing can be good at times during fall, when large brown trout work their way upstream from Otter Creek.

Sugar Hill Reservoir (Goshen Dam)

If fishing the Neshobe River in spring, particularly as far up as the North Branch, you'll want to take a drive up to neighboring Sugar Hill Reservoir, also called Goshen Dam. You'll need a small floating craft to effectively fish the water during late April and May, when VF&WD stocks more than 3,000 brook trout that range in size from 8 to 15 inches. It's known as one of the most popular trout fishing destinations in Green Mountain National Forest because of its easy access via a formal boat launch at the dam.

From the VT 73/US 7 intersection in Brandon, take VT 73 past Forest Dale about 5 miles to a left turn on Town Hill Road (FR 32). In less than a mile, take a left turn on Carlisle Hill Road (still FR 32), where you'll cross the North Branch of Neshobe River. Continue on FR 32 past Goshen Town Hall. (FR 32 is also named Goshen-Ripton Road once you're up that far.) In a little over 3 miles from Goshen Town Hall, take a right turn on Goshen Dam Road at the intersection with FR 32. Signage at the intersection points the direction to Sugar Hill Reservoir. Goshen Dam and the boat launch are about a mile up Goshen Dam Road.

During very early season, the gate at the entrance to Goshen Dam Road might be closed until the gravel road dries out a bit. Parking and unloading your watercraft is easy. Make sure you sign in at the parking area. During late April into May, you'll encounter conventional tackle anglers both on the water and casting from the fishing platform at the launch. Fall is a great time of year to row or paddle around Sugar Hill Reservoir.

I asked VF&WD fisheries biologist Sean Good what the likelihood is for holdover brook trout beyond the stocking year. "Before GMP [Green Mountain Power] drained it completely in 2021, we did a trout survey and did not find any holdover brook trout from previous year's stockings. We did find a good number of large wild brown trout—a species

Sugar Hill Reservoir, also called Goshen Dam, is located a short distance from the Neshobe River. It's primarily a put-and-take stocked brook trout water but also supports brown trout. A fishing platform popular with conventional tackle anglers is located at the launch site. VALERIE VALLA

we don't stock in there—which likely migrated into the reservoir from Sucker Brook. Some of these exceeded 20 inches," Shawn said. Green Mountain Power, which operates the dam, drained the reservoir to complete dam safety enhancements on the 58-acre reservoir. Brook trout stockings resumed in May 2023.

Cold River

Flowing out of the Coolidge Range in the Green Mountains near North Shrewsbury, Cold River traverses nearly 13 miles before entering Otter Creek just north of the town of Rutland. The stream is stocked with both brown and brook trout. About 1,000 brook trout are stocked annually from North Shrewsbury downstream to Upper Cold River Road. During May, an equal number of browns are stocked from the Upper Cold River Road area downstream to Middle Road in Clarendon.

Access the stream by taking Cold River Road off US 7 in Rutland. You'll intersect Upper Cold River Road in about 4 miles. Have a look at the river at the Browns Covered Bridge, located just a short distance from the Cold River Road/Upper Cold River Road intersection. This reach is known not only for trout water but also, probably more importantly, for the charming historic Brown Covered Bridge that spans Cold River just 0.2 mile from the intersection.

Park at the bridge, where you'll find a footpath that leads down a slope to the water. Keep in mind that Cold River is just that during summer—cold water that attracts folks looking to sit in the water downstream from the bridge. Head downstream from the bridge, fishing pocketwater and small pools along the way. It's nice to wet-wade the stream during summer's heat. I fish a 6-foot, 3-weight rod during late summer.

Cold River at historic Brown Bridge on Upper Cold River Road. By summer the little stream runs low, clear yet cold. However, the more favorable time to fish the stream is during late May. VALERIE VALLA

Before departing the Upper Cold Brook Road reach, take a few moments to read the State of Vermont historical marker at the covered bridge. While about 100 covered bridges are scattered throughout Vermont, Brown Covered Bridge, built in 1880, is especially significant. It was formally designated a National Historic Landmark by the National Park Service in 2014.

A second stretch that provides easy access to the water is a few miles upstream. It's best to return to Cold Spring Road and then drive about 3 miles or so to the Wilmouth Road crossing. The stream is directly next to the road in that area. Once the stream departs the crossing, it eventually flows away from the road and out of sight.

Mill River

Mill River, located not all that far from Cold River, has seen better days. Like so many trout tributaries around Vermont, including Cold River, Mill River has been hammered and changed by significant high rain events. Tropical Storm Irene in 2011 and a two-day heavy rain that struck Vermont in July 2023 rushed high and powerful flows down from the mountains. The storms ripped away roads and riverbanks, scattering fields of boulders and rocks along the streams.

Mill River, located 10 minutes or so from Brown Bridge on Cold River via East Clarendon Road and VT 103, starts its 17.8-mile journey to Otter Creek, flowing out of the Green Mountains at the town of Mount Holly. VF&WD stocks the river with both brook and rainbow trout annually. About 1,000 yearling brook trout and 500 rainbows are stocked between Barlow Road bridge (off VT 155) in the Mount Holly area and Spring Lake Road in Shrewsbury (off VT 103). The river courses its way along VT 115 until it reaches East Wallingford; it then flows along VT 103 past Cuttingsville.

Mill River along VT 103 next to Spring Lake Road in the Shrewsbury area. Spring Lake Road marks the lower boundary of the stocked section. During late season, the stream runs low and clear.

You can get into the water at a VT 103 bridge crossing next to Spring Lake Road, a few miles downstream from Cuttingsville. If you're anticipating fishing this reach, understand that summer isn't a great time to cast your lines. The river shallows along the fields of washed stone caused by floods. As is true of many of Vermont's waters, late May into early June is prime time. You might pick up fish later in the season at bankside along areas where the river narrows. During summer you would be better off fishing the river's headwaters.

Most visitors to Mill River are interested in the Clarendon Gorge area, a popular section just upstream from where Mill River passes beneath US 7. Hikers traversing the Appalachian Trail trek through this beautiful gorge. A popular viewing area of the gorge can be accessed at the Clarendon Gorge Appalachian Trailhead parking turnout on VT 103, about 2.5 miles beyond the bridge crossing at Spring Lake Road. The trail leads to the Bob Brugmann Suspension Bridge.

If fishing is on your mind, and not simply viewing the gorge from the suspension bridge, head to Pluss Lane off VT 103 just beyond the trailhead lot. Take Pluss Lane to the intersection of Airport and Gorge Roads. Access the stream off Gorge Road by carefully parking along a safe area on the road. Look for a footpath that leads to the water. If approaching from US 7, park at a point 0.3 mile from the US 7/Gorge Road intersection. Keep in mind that the river here can be dangerous after high rainfall events, with floodwaters rushing down through the lower gorge. During the heat of summer, swimmers will occupy the area. There's no telling what fish species you might run into along the area, marked by colorful marble and basalt rock. Rainbow trout occupy the lower reach at times, as well as other warmwater fish. It's worth a visit, fish or no fish.

Middlebury River

Larger, more substantial Otter Creek tributaries that support trout are located downstream from Rutland. Chief among them is the Middlebury River, and not all that far from Middlebury River is the New Haven River. Born in Vermont's Green Mountain National Forest, 18.6-mile-long Middlebury River brings together its middle and south branches (two of three branches) at Ripton. The middle and south branches combine before entering Middlebury Gorge, a magnificent natural area popular with kayakers who enjoy negotiating its Class IV–V waters and summer swimmers attracted to its cool waters.

After exiting the gorge, the Middlebury River flows more gently through lower elevation flat terrain past East Middlebury. The stream drops from an elevation of 2,400 feet in the Green Mountain National Forest headwaters to 500 feet once it emerges from the gorge. It meanders through agricultural areas along the Champlain Valley flatlands then dumps into Otter Creek not far from East Middlebury.

Middlebury River at the VT 125 bridge crossing at the North Branch Road intersection. You'll encounter swimmers and other users along this section during the heat of summer. VALERIE VALLA

Juggernaut 2.0 *Created and tied by Rich Strolis*

HEADWATER AREA AT RIPTON TO THE GORGE

If you're after brook trout, head to the Middlebury River branches. A 6.5-mile drive up VT 125 from US 7 to the small town of Ripton will put you in range of pleasant headwater fishing experiences. There's not much to Ripton aside from schools and the Ripton Country Store, where you can pick up deli items and drinks for your outing. VF&WD stocks a few hundred brook trout on the South Branch from about Brooks Road to Ripton. On its way to Ripton, the South Branch picks up water from two small tributaries, Brandy Brook and Goshen Brook. I've picked up small brookies on Goshen Brook, about a mile upstream from the Goshen Road/VT 125 intersection.

A short distance up Goshen Road, there's a road shoulder area where you can get into the tiny stream. Continue up Goshen Road less than a mile, just beyond the Spirit in Nature Interfaith Nature Path parking lot. You'll encounter a large "Moosalamoo National Recreation Area" sign at an intersection with FR 91, a mountain bike route. The stream flows under a small bridge at the intersection. This is very small water, and at times water levels are low. I use a 6-foot, 3-weight Becker Cane along this reach when water levels are optimum. Goshen Brook feeds the South Branch a short distance downstream. You'll also want to check out the South Branch a short distance upstream.

Continue on VT 125 a little over a mile from the VT 125/Goshen Road intersection (5.8 miles from the VT 125/US 7 intersection). You'll encounter the Robert Frost Wayside picnic area on the left and a trail loop on the right. The National Recreation Trail Loop commemorates Robert Frost's poetry. The first 0.3 mile is an easy trek along a boardwalk that passes over beaver pond areas out to the South Branch.

VF&WD also stocks a few hundred brook trout in headwater areas of the Middle Branch from Wagon Wheel Road downstream to the gorge and on the North Branch from the uppermost Lincoln Road crossing downstream to Dugway Road.

GORGE AREA TO OTTER CREEK

If you drove to Ripton along VT 125 from US 7, as you climbed up in elevation, you surely noticed the guardrail next to the road the entire way to the town. This makes for difficult parking access, even if you intended to scale down the steep hill to the gorge. One slight road shoulder pullout could accommodate a vehicle—for those crazy enough to have a look. The North Branch comes in at river right, down in the gorge, the combined branches now flowing as one.

While some adventurist anglers might contemplate sampling the gorge section, gorge fishing ventures are best left to those who have experience negotiating such areas. I can't even recommend it, although I know some fly anglers might smile at that notion. Inherent dangers with such ventures can't entice me to get down the hill, no matter what fish could be brought to the net. In my younger years, I surely would have tried venturing down into the gorge in search of trout, but that is not one of my desires now.

Steven Atocha, owner of Middlebury Mountaineer sport shop and a highly experienced guide, doesn't recommend that adventure: "We do not guide the Middlebury Gorge due to how dangerous it can be in there; because of all the flooding in recent years, the footing can be sketchy. We always recommend that anglers enter from the Route 125 bridge and not try to scale down from the upper section. The access is okay from Ripton down to the junction with the North Branch of the Middlebury, but due to a waterfall and cliffs below this section, there is no real way to continue further downstream," Steven said.

A nice place to view and reasonably fish the lower gorge, for at least a short distance, is off the VT 125 bridge crossing next to the intersection with North Branch Road. During the heat of summer, you'll likely encounter many vehicles at the road shoulder parking next to the bridge. You can access the stream directly below the bridge with little issue. Park on the far side of the bridge then walk across the bridge to a small crushed-stone path that quickly leads to the river and a large, slow-moving pool. You'll come across interesting signage near the footpath that alerts visitors about the East Middlebury Ironworks that existed at the site between 1831 and 1890. Remains of the rock foundation are next to the stream, shrouded by trees. Sampling areas downstream in the gorge itself is a significant issue. If after stocked rainbow and brown trout, head downstream from the gorge. About 600 rainbows and 400 browns are stocked from VT 125 below the gorge to Otter Creek.

Steven Atocha's shop, conveniently located on 56 Main Street in Middlebury, can fix you up with all the flies you should carry while fishing Middlebury River. Steven and his wife, Marion, started Middlebury Mountaineer in 1998 and have been guiding wade trips as well as float trips since then. The half dozen guides working for them fish the following waters: the Lamoille, Winooski, New Haven, Middlebury, Neshobe, and White Rivers, Lewis and Otter Creeks, and most of the tributaries that run into those rivers.

Among the flies he recommends are Frenchie, Walts Worm, Prince, RS-2, Cornfed Caddis, Red Dart, Stimulator, Pheasant Tail, Hi Vis Parachute Adams, assorted Hoppers, Hi Vis Parachute Ant, Girdlebug, assorted Perdigons, Blowtorch, Quilldigon, Tan Caddis, and Noonies Purple Pleasure. The shop carries a wide selection of dry flies, nymphs, and about any pattern you might want.

Of course you can pack your own favorites for the hatches Steven says you might encounter on the Middlebury River. During April into May you'll want your boxes to include March Browns and Hendricksons. During May and June, Yellow Sallies appear; late spring into early summer, *Isonychias* pop up. During summer, Golden Stoneflies, Flying Ants, and Hoppers work well. Blue-Winged Olives are around anytime the water is around 50+°F. A larger *Isonychia* hatch shows up in September. October Caddis appear during October.

Lewis Creek

STEELHEAD REACHES

A quick 15-minute drive north from Vergennes via VT 22A/US 7 North will get you to Lewis Creek, another Lake Champlain tributary in the area. Lewis Creek, like Otter Creek, is a stream that supports two fisheries. Stream reaches downstream from the falls in North Ferrisburgh are best characterized as a warm fishery, although fly fishers target steelhead that run up from Lake Champlain. You'll have plenty of company around the falls, especially in early spring. You'll want to target steelhead during March and April, depending on year-to-year seasonal changes. Later in the season, warmwater species congregate near the falls. The falls are located upstream from a small bridge crossing on Old Hollow Road (the first right turn after the US 7 bridge crossing if driving north). Anglers also access Lewis Creek downstream from the US 7 bridge crossing, just south of North Ferrisburgh. On the north side of the bridge, take Loven Lane a short distance to a dead end at a guardrail. A footpath on the other side of the guardrail will take you down a short slope to the stream. A variety of fly patterns will get the attention of the steelies—everything from Woolly Buggers to nymphs. Depending on water levels and conditions, a sink tip line is handy to get your fly down. I use a 9-foot, 6-weight rod.

Every steelhead water I've ever fished seems to have one or two individuals who far exceed others when it comes to understanding the system's history, its nuances, and the

Head to North Ferrisburgh Falls with hope of netting a steelhead during late winter and very early spring. During May, smallmouth bass congregate in the area.

best fishing tactics that can lead to success in netting a nice fish. I doubt there's a more experienced and knowledgeable Lewis Creek steelheader than Vermonter Ralph Kucharek. Ralph operates The Vermont Fly Fishing Company out of nearby Burlington. Without hesitation, I'd recommend Ralph to any fly fisher interested in sampling Lewis Creek's steelhead fishery. He also guides other area rivers and waters, targeting other fish species. Ralph kindly provided the insight I was looking for surrounding Lewis Creek and its steelhead, including great background information.

"We had a steelhead program that started in 1970s," Ralph said during a nice chat. "At first, Memphremagog-strain steelhead from Lake Memphremagog in the eastern part of Vermont were introduced into the Lewis Creek steelhead system. That strain is typically a spring-run steelhead, and only a handful end up in Willoughby River and other tributaries during the fall." In the 2000s Vermont started playing around with other strains after catch-rate surveys were completed, studies that involved Ralph and other knowledgeable steelhead anglers. VF&WD eventually switched to Chambers-strain steelhead, the same strain that's stocked on Salmon River in New York. Chambers strain worked out better, for a few reasons.

Ralph, who typically targets March 1 for steelhead guiding on Lewis Creek, has worked out tactics that have brought his clients successful outings. "We target Lewis Creek steelhead a couple of different ways," he explained. "The most effective way is nymphing, using Indicators. We've found the most effective flies are White Woolly Buggers. My theory is that the spring runoff can be pretty murky. White provides contrast that shows up well in the water."

Ralph prefers unweighted flies, utilizing split shot to get the fly down where he wants it. "Chartreuse or pink Estaz Eggs and a rubber-legged Chris Crack Estaz Egg work well too." The steelhead are not extremely fussy. Other factors besides fly selection can lead to a successful outing over a blank day. "Confidence plays a big role in successful steelheading," he added.

While some fly anglers fish double rigs, Ralph prefers fishing only one fly, on-point, for a couple of reasons. "There's always a chance of foul hooking a fish, and general presentation works better for me by fishing a single fly." Ralph also provides useful fishing tactics, not limited to rigs, while guiding clients on Lewis Creek. Steelheading on Lewis Creek, like other steelhead waters, requires understanding where the fish might hold—and how water levels and conditions require a change of tactics. An important part of Ralph's guiding is educating his clients while on water.

Tied by Mike Valla

Chartreuse Egg patterns and White Woolly Buggers are flies you should have in your box while steelheading on Lewis Creek.

Pink UV Estaz Egg, Chartreuse Estaz Egg, Chris Crack UV Estaz *Tied by Ralph Kucharek*

"I explain that the steelhead generally like 'walking speed water' and pockets, but the fish aren't always going to be located in what would appear to be the most likely holding areas. I tell clients to not throw their first cast by even stepping in the water; steelhead could be holding right at their toes, close to the bank. Another thing I tell clients is to always fish out your drift to the end. I've taken fish at the point my fly has swung to the bank, and that's where they hit. On Lewis Creek, don't assume where in the stream you'll get a fish—fish it all."

Another important variable Ralph pays attention to is water temperature. He looks for temperatures around at least 42°F. "Once you get below that temperature, the fish get lethargic. They head for deeper pools, where the fish are during low temperatures. That's where they will expend the least amount of energy. The steelhead that were hunkered down in the pools at low water temps will vacate if the water suddenly climbs a few degrees, to say 50°F. They'll move up into fast water at that point," Ralph said. The moral: Carry a thermometer. Ralph believes that water temperature can be more important than water clarity. He's netted steelhead when the water was like chocolate milk. He added that Lewis Creek fishes best under 250 cfs (cubic feet per second). "I look for bumps in water levels too. If the water is coming up, you'll always want to be fishing on the drop. Steelhead, in my experience on Lewis Creek, don't like rapid water-level changes. Be aware if the stream levels are coming down."

The best time of day to get on water is another factor Ralph has studied—the very early mornings so often believed to be the best on many steelhead streams are not always the most optimum times to fish in the day. Fishing could be much better during the afternoon on a day when higher ambient temperatures are reached after a series of cold days. "Pay attention to those kinds of optimum windows on Lewis Creek," he suggested.

With the amount of time Ralph has been fishing Lewis Creek, he has no doubt had many interesting encounters during steelheading. One such story that came to his mind while we were chatting about Lewis Creek steelhead involved a particular catch. "I caught a tagged steelhead, which I thought was kind of cool. I submitted the tag to VF&WD and received back a report that indicated the fish's size, weight, and other details when the fish was tagged. It turned out the fish had been tagged three weeks prior to my netting it, 30 miles north on Lake Champlain. Those fish are migratory and always moving," Ralph said.

You can get into fish downstream from Loven Lane, where steelies sometimes hang out before heading downstream back to Lake Champlain. But Ralph made a point that the section between Loven Lane upstream to the falls is the "bread-and-butter" reach for steelhead.

UPPER TROUT REACHES

Sections upstream from the falls support rainbow, brook, and brown trout and sometimes steelhead that make it above the falls. Lewis Creek is accentuated by quaint covered bridges that span the stream along its upper reaches. In its upper Monkton/Hinesburg area, Lewis Creek flows through a Streambank Management Area (SMA). The VF&WD purchased the lands to ensure access for anglers and other recreational users and to address habitat improvement needs for aquatic species.

If you're targeting trout and seeking an experience in the Lewis Creek headwaters SMA, formal VF&WD parking at Lewis Creek is available at a crossing on Silver Street in Hinesburg. It's located about 10 miles from the US 7/Monkton Road intersection (just outside of Vergennes). Take Monkton Road to a left on Monkton Ridge then a slight right onto Silver Street. Another parking area, located upstream in a slow-moving section more appropriate for small craft such as canoes, is about 3 miles from Monkton Ridge. Drive east on State Prison Hollow Road (near Alderman's of Vermont Chocolate Shop) to a left on Tyler Bridge Road. You'll see a formal VF&WD "Lewis Creek at Tyler Bridge Access" sign next to the small bridge crossing.

Seguin Covered Bridge reach on Lewis Creek provides scenic fly fishing along nice runs and stretches inhabited by brook and brown trout. Fishing alongside a quintessential Vermont covered bridge makes for an enjoyable session, fish or no fish. VF&WD typically stocks about 1,000 brown trout and 1,000 brook trout in its upper reaches.

A couple other more pleasant reaches for fly fishing are located downstream from the SMA at Seguin Covered Bridge on Roscoe Road and another downstream at Quinlans Covered Bridge where Monkton Road, Spear Street, and Lewis Creek Road converge. The Sequin Covered Bridge reach is particularly picturesque, with the stream plunging over a drop just upstream from the bridge. A small road shoulder parking spot is located on the far side of the bridge. To reach Quinlans Covered Bridge from Seguin Covered Bridge, drive right on Roscoe Road about a mile to a right turn on Lewis Creek Road. You'll reach Quinlans in about 1.6 miles on Lewis Creek Road.

New Haven River

New Haven River's nearly 60 miles of water (if you count its extreme headwater sources) begins from small springhead flows that coalesce deep in the Green Mountain National Forest near the Breadloaf Wilderness. Its sister stream, Middle Branch of Middlebury River, is born not all that far away. Flowing at first as a small brook, the little stream departs its birthplace at an over 2,500-foot elevation. Weary backpackers on extended wilderness trips enjoy its cool, soothing riffles, runs, and small pools along several trails along its headwaters. Hikers that head to the Cooley Glen/Emily Proctor trailhead on South Lincoln Road surely admire the stream at points where it comes into view.

South Lincoln Road eventually follows New Haven more closely as the stream pushes toward the town of Lincoln. The little stream can be viewed at the Grimes Road bridge crossing off South Lincoln Road, not far upstream before South Lincoln Road ends at a T intersection. From the bridge you can watch the little stream flow through a tumbling gorge. It would take a pretty brave and ambitious fly angler to consider scaling down into

New Haven River along River Road, several miles downstream from Bristol, flows through good structure. While stretches on the lower reaches are open to public fishing, be aware of posted property signs. Always respect private property. VALERIE VALLA

that gorge. One can only imagine the brook trout that surely inhabit that reach. Fly anglers should head to more accessible waters downstream from Lincoln. Access to the stream along South Lincoln Road upstream from the T is difficult, aside from one or two spots directly upstream from the intersection.

At the T intersection where South Lincoln Road ends, seasonal Lincoln Gap Road, which leads over the mountain to the Mad River Valley, continues to the right; East River Road continues to the left, where it enters Lincoln then quickly departs as East River Road. A couple of access points are available at roadside shoulders a mile or so downstream from Lincoln.

Lincoln is located near the foot of Mount Abraham. Incidentally, although the town wasn't named for our 16th president, the mountain was indeed named for Abraham Lincoln.

After exiting the Lincoln area, now as Lincoln Road, the stream pushes toward Bristol. Along the road, you'll pass Eagle Park about a mile before Lincoln Road intersects VT 116. Eagle Park has a parking lot, picnic tables, and an ADA-compliant universal fishing platform—one of the first riverside UFPs located in Vermont. The Eagle Park site provides some of the easiest access to the river before entering Bristol.

On its way to Bristol along Lincoln Road, not far downstream from Eagle Park, New Haven River drops through marble formations that create 16-foot-high Bartlett Falls. One of the most popular swimming holes in the Northeast, it's located shortly before Lincoln Road ends at the VT 116 West/VT 17 East intersection, just outside Bristol. You will want to bypass fishing this area during the heat of summer. You'll take a left turn on VT 116 to get to other nice fly-fishing sections in and around Bristol.

Just before entering Bristol, driving on east on VT 116, you'll see the historic Bristol Rock next to a nice streamside picnic area at a large parking lot, complete with a couple of tables where you can have your sandwich. Dr. Joseph C. Greene paid a carver to chisel the Lord's Prayer into the large triangular rock in 1891. Access into the stream is quick and easy. If you're fishing the stream reach during late May into June, you'll likely latch into a rainbow trout along the runs both upstream and downstream from the picnic area. Bristol has excellent fly-fishing sections, including a stretch along Lower Notch Road in town. My wife, Valerie, and I had an opportunity to rent a small streamside cabin on Lower Notch Road, where we enjoyed casting our small streamers and nymphs for rainbows.

Access is also available downstream at Sycamore Park, located off VT 116 a couple of miles south of Bristol. The large parking area is at a VT 116 bridge crossing about 0.2 mile from the VT 116/River Road intersection. Once the river leaves Bristol, it calms down a bit and begins meandering through the broader Otter Creek watershed valley.

If interested in sampling its lowermost reaches continue south on River Road rather than VT 116. You'll come across some areas that are posted along the river, but you can get into the river at a River Road shoulder turnout about 4 miles or so south of the VT 116/River Road intersection (3.6 miles north of US 7 if approaching from that direction). DeMeres Park is another nice access area on River Road, about 1.3 miles from its intersection with US 7. A gravel road leads down a slope to the parking and picnic area. A footpath from the parking area leads to the stream. The New Haven River continues its journey toward Otter Creek, flowing through farm country and cornfields. Its final drop tumbles about 6 feet over the sheer ledges of Dog Team Falls, just upstream from its confluence with Otter Creek.

The falls were named for Dog Team Tavern; built in the 1920s, it was destroyed by fire in 2006. I had the opportunity to rent a streamside place directly at the falls. It's a gorgeous piece of water, popular with anglers. The falls are located on Dog Team Road, about 0.5 mile from the Dog Team Road/US 7 intersection, if you're approaching from

Dog Team Falls on the lower New Haven River is not far upstream from the river's confluence with Otter Creek.

the south. Be careful where you park, since this is a residential area. Most anglers park at a wide gravel road shoulder a few hundred feet beyond the bridge crossing.

The New Haven River receives an ample number of stocked fish annually. From its confluence with Otter Creek to River Road and South Bridge Street to Bartlett Falls, Vermont stocks about 2,000 rainbows and 2,000 brown trout. From Bartlett Falls to Lincoln Gap Road, the stream gets about 700 to 800 brook trout. About 4 miles of Otter Creek, from Munger Street Bridge in New Haven to South Street Bridge in Bristol, are under Special Regulations. A protected slot harvest size (10 to 16 inches) is in effect—of little consequence to catch-and-release fly anglers.

Western Coachman *Tied by Mike Valla*

Western Coachman is a simple pattern, first fished on Tule River many years ago. The fly was designed as neither a wet nor a dry fly but rather as a semi-dry fly that floats just beneath the surface. It was first field-tested for rainbow trout and proved to be very effective, eliciting interest from that species.

BALDWIN CREEK TRIBUTARY

While in the Bristol area, take a quick run up to one of New Haven River's little tributaries, Baldwin Creek. From the VT 116/VT 17 intersection with Lincoln Road, drive up VT 116 North/VT 17 East less than 2 miles to the VT 17 East intersection with Drake Woods Road. Take a look at the creek at the bridge crossing, then continue on VT 17 East (Drake Woods Road) about 1.6 miles to 19-acre Bristol Memorial Park, just before the bridge crossing.

Just after the bridge crossing, you can also access the stream at the Edwin Wright Orvis Memorial Park at the Bristol–Starksboro town line. You'll want to trek down a path that leads from the small parking area at Bristol Memorial Park to small, scenic Burnham Falls. Baldwin Creek is accessed by footpaths that lead from the parking area.

The forested lands along the stream where you can fish for brook trout were procured by the town in the 1950s. The Bristol Selectboard, the Bristol Conservation Commission, and the Bristol Recreation Department manage the lands. I like to fish my 7-foot, 3-weight rod, plopping small bead-head patterns and small Elk Hair Stone dry flies along the small little pools and riffles. In late summer, Baldwin Creek's cool waters are perfect for wet-wading or wearing hip boots. Vermont stocks about 600 brook trout along the stretch from Russell Young Road to VT 116.

While fishing the upper sections of New Haven River in the Champlain Valley, you'll want to consider a drive over Lincoln Gap Road to Mad River and the Mad River Valley. Again, keep in mind the road is seasonal; the upper portion typically closes from mid-October to mid-May (adjusted for conditions). You can check the road open status at: https://lincolnvermont.org (under "Community Town News"); the town also posts a large orange sign that alerts travelers if the road is closed. The towns of Lincoln and Warren (in the Mad River Valley) work together to determine exactly when the road will be blocked with cement barriers and "Road Closed" signage.

Winooski River

One of Vermont's longest rivers, the Winooski flows 90 miles from its birthplace in the town of Cabot westerly to the town of Colchester just north of Burlington, where it feeds Lake Champlain. Parker Wright, who has been guiding fly fishers out of The Fly Rod Shop in Stowe for many years, provided a few insights based on his many outings on the Winooski and its tributaries. Parker hails from Waterbury, just north of Stowe.

Samantha Kearney on the run upstream from the Winooski Street bridge crossing, off River Road, during lower water conditions in early October. The crossing leads into Waterbury. Her fishing companions, Cole Perkins and Scott Smullen, are working water downstream near the bridge.

Chloe Benoit landed this fine brown during a March outing on one of the Winooski's tributaries. Winooski's tributaries are as interesting as the river itself. PARKER WRIGHT

"The Winooski tends to fish best from mid-May to mid- to late June and picks up again in early September through mid-October. More fish are caught nymphing, although there are pockets of exceptional dry fly activity during late May, June, and September. The Winooski is stocked every spring, although we run into plenty of wild and holdover fish ranging in sizes from 6 inches all the way up to above 20 inches. Some of the larger tributaries of the Winooski hold some incredible wild browns that can rival fish caught throughout the country." If making a trip to the river, stop by The Fly Rod Shop for great advice, gear, and flies. Parker and the team at the shop will get you into fish on area waters.

At first the river flows southwesterly from Cabot to Montpelier; then, in a "V" form, it changes direction and flows from Montpelier northwesterly to its mouth at Lake Champlain. Along its course, the river system is fed by many small tributaries that serve as spawning and nursery waters for brook, brown, and rainbow trout. Anyone interested in a better understanding of the upper Winooski system in all of its complexity should read VF&WD fisheries biologist Bret Ladago's 2017 *Upper Winooski River Fisheries Assessment*. It's available online. Bret's excellent report breaks down the system from its headwaters in Cabot to the top of Bolton Dam in Druxbury. All the small feeder brooks are listed, along with a breakdown of the main river into logical segments from the source area to Bolton Dam.

A few details about the headwater reaches will be addressed, but the bulk of attention here is focused on the Winooski's more popular fly-fishing stream reaches and access points along its middle section, downstream from Middlesex. Details of the landlocked salmon migration that occurs during the fall out of Lake Champlain at Burlington will also receive some attention. Most anyone familiar with fly-fishing the Winooski would note the interesting major tributary fishing opportunities. Seven major tributaries feed the Winooski River. Entering the river from the south are the Mad, Dog, and Huntington Rivers and Stevens Branch. Entering from the north are the Little River, North Branch, and Kingsbury Branch. Of the seven, the Dog, Mad, Little, and Huntington Rivers are among the more popular with most fly anglers. Those four streams will follow this Winooski River section.

VF&WD has identified a number of upper Winooski watershed tributaries that are important spawning and nursery habitat waters. The following waters were listed in the *2017 Basin 8–Winooski River Watershed Water Quality and Aquatic Habitat Assessment Report*, prepared by Vermont Agency of Natural Resources and Department of Environmental Conservation:

- Winooski River (above Cabot Village): wild brook trout
- Molly's Brook (above Marshfield Dam): wild brook trout and brown trout
- Kidders (Hooker) Brook: wild brook trout and brown trout
- Jug Brook: wild brook trout
- Beaver Meadow Brook: wild brook trout
- Nasmith Brook: wild brook trout, brown and rainbow trout; important spawning tributary for main stem populations above Plainfield Dam
- Great Brook: wild brook trout, brown and rainbow trout; important spawning tributary for main stem populations above Plainfield Dam
- Sodom Pond Brook (tributary to Kingsbury Branch): important spawning tributary for main stem populations
- Dugar Brook (Kingsbury Branch tributary): wild brook trout
- Hancock Brook (tributary to North Branch): wild brook trout
- Minister Brook (tributary to North Branch): wild brook trout
- Martins Brook (tributary to North Branch): wild brook trout and brown trout
- Patterson Brook (tributary to Martins Brook): wild brook trout and brown trout
- Herrick Brook (tributary to Martins Brook): wild brook trout and brown trout
- Jones Brook: wild rainbow and brook trout; important spawning tributary for main stem trout population
- Crossett Brook: wild brook trout, brown and rainbow trout; important spawning tributary for main stem trout population
- Thatcher Brook: wild brook trout, brown and rainbow trout; important spawning tributary for main stem trout population

HEADWATERS TO PLAINFIELD

VF&WD surveys have shown brook trout to be abundant upstream from Cabot Village, where they inhabit small flows that combine to form the Winooski after departing Coit's Pond. Downstream from Cabot, the tiny river flows at first through a degraded area then through farmlands as it roughly follows VT 215. Riparian cover is intermittent as the stream pushes toward the McCrillis Road/VT 215 intersection. The stream continues along VT 215 to Marshfield, where VT 215 intersects US 2. As occurred along other Vermont streams such as the Lamoille River, as you drive along the river you'll notice the effects of the long July 2023 rain event that resulted in catastrophic floods. Trees were uprooted and washed into the streams. Sand and silt covered riverside fields and roads were washed out during that historic natural event.

In Marshfield you can park next the US 2 bridge crossing next to the Wooster Cemetery. It's across the road from the Marshfield Volunteer Fire Department. Some good riffle water can be found in the Marshfield area but opens up again to unremarkable open and exposed areas. The Winooski slows as it approaches the dam at Plainfield. In their survey work a few years ago, VF&WD found a mix of stream born rainbows along with a few wild browns. However, about 1,200 hatchery rainbows are stocked annually between the confluence of Molly's Brook near Marshfield downstream to Plainfield Dam area. VF&WD Department owns three parcels of land along the Winooski River around Plainfield and Marshfield, which provides angler access and riparian habitat.

Fly fishing is possible here and there as the river snakes back and forth along its 10 miles through agricultural lands and settlements between Marshfield and Plainfield. The same can be said about a few stretches as it approaches Montpelier downstream from Plainfield. Drive along US 2 west of Plainfield and get into the water at points that appear to be good trout holding water. While a few wild trout are in the mix, you'll likely be targeting hatchery rainbow trout. The stretch where Town Hill Road intersects US 2 has some decent stream structure. However, river water temperatures can reach high levels during summer so you'll want to fish that reach during late April into late May for best results.

While you'll encounter plenty of conventional tackle anglers around Montpelier, and at the mouth of Dog River that joins the Winooski just west of the city, especially during May after the river has received stockings, some of the better fly-fishing water happens starting at Middlesex, where wade fishing is possible along nice water that has more structure than the flat, slow-moving water in the Montpelier area. Conventional tackle anglers sometimes park at the Dog River Recreation Center, on Dog River Road in Montpelier, where they can get a canoe or small craft into the Dog River and float down to the Winooski.

MIDDLESEX-TO-WATERBURY SECTION

Mad River joins the Winooski River off US 2 in Middlesex. The dam reach upstream from the confluence of both waters can be good, especially during spring and early fall. Good water is found along areas between Middlesex and through the Waterbury area. Anglers get into a mix of hatchery trout and a few wild trout at bridge crossings in Waterbury. The US 2 bridge crossing in Waterbury, where Crossett Brook joins the river, is popular with float anglers who put small watercraft into the water at the bridge. Parking is available at downstream side of the bridge.

VF&WD fisheries surveys have shown Crossett Brook is inhabited by wild rainbow trout and wild brown trout. The brook is an important spawning tributary for Winooski River trout. Like so many other large, wide, deep-pooled, and often seasonally thermally challenged rivers of this nature, not only in Vermont but anywhere a brook (no matter how small) joins a big main stem river, the confluence can often be a likely holding, refuge, or staging area for trout during certain times of the year. This is especially true during spring and fall. The same holds true for other areas along the Winooski where trout-inhabiting tributaries enter the river.

SPECIAL REGULATIONS TROPHY TROUT SECTION

From the US 2 bridge crossing along Crossett Brook, the Winooski flows just over a mile downstream to the historic Winooski Street truss bridge. The bridge, built in 1928 after catastrophic 1927 floods wiped out nearly 1,200 bridges and even small communities in Vermont, is listed in the US National Register of Historic Places. Beyond its significance from a historical perspective, the Winooski Street bridge crosses the Special Regulations Trophy Trout stretch that runs around 5 miles from the US 2 bridge crossing east of

Waterbury Village downstream to the top of Bolton Dam (Bolton Falls). It involves a lower portion of the Little River tributary.

Of significance to the river, and fly fishing, Little River pours its cold water into the Winooski about 1.5 miles downstream from the Winooski Street bridge. Little River (discussed in detail in later sections) exits Waterbury Reservoir as a tailrace barely 3 miles upstream. Cool water entering the system provides excellent relief to rising water temperatures during summer. Large holdover browns that seek spawning areas swim into Little River during fall. You can access the mouth of Little River by taking Farr Road off US 2, just west of Waterbury. A parking area is located with VF&WD signage next to the overpassing highway.

BOLTON FALLS TO RICHMOND SECTION

You'll want to locate to River Road on the south side of the Winooski to access some of the really nice fly-fishing runs and pocketwater. At first, as you're driving down River Road from its intersection with the Winooski Street bridge, the river vanishes from sight. In less than a mile, the river comes back into view. Be aware that in a little more than 0.5 mile from the intersection, River Road becomes a dirt road, leaving pavement behind. It's not an issue really once the road has been spiffed up and regraded by highway crews after the long winter. However, I've driven that dirt section, which runs for several miles, during very early spring before the road has been repaired, and more deep, often-muddy ruts and potholes spread across the width of the road than I've ever encountered elsewhere. Yet good fly-fishing water along River Road makes the aggravation well worth it.

Stretches downstream from Bolton Falls off River Road are undoubtedly some of the most popular fly-fishing stretches on the Winooski River. The DeForge Hydroelectric Station Recreation Area at the dam provides easy river access. There's a large parking lot next to the river at the dam, down Power Plant Road off River Road.

When water levels allow easy wading, especially during cooler early fall days, try the run near the Winooski Street bridge crossing off River Road that leads into Waterbury. In spring, water levels along this stretch are often much higher and difficult to wade. During wade-friendly conditions, park on the Waterbury side of the bridge, then slip down a bank slope on the upstream side of the bridge to access the water.

About 4 miles from the Winooski Street/River Road intersection (2 miles beyond where River Road's pavement ends and the dirt begins), you'll begin to see road shoulder parking spots and enter stretches in the Bolton Falls reach. There's no doubt the section downstream from Bolton Falls is among the most popular fly-fishing stretches on the river. Abundant rainbow trout are easily caught during May after seasonal stockings downstream and along the railroad bridge crossing, a short distance downstream from Bolton Falls. After spring runoff levels drop river volume, this section is much easier to fly fish. Woolly Buggers work well along the runs that become much more accessible during May.

There's an access point very close to the Camel's Hump Road/River Road intersection along Ridley Brook, an important Winooski tributary. A small parking area is located next to the brook that leads to riffles and runs after it flows through a box culvert under the railroad tracks. You'll see VF&WD signage posted along trees that alerts anglers that Ridley Brook is a Special Regulations stream. The brook is closed to fishing the second Saturday in April through May 31. Slot size limits are in effect.

A short distance downstream from the confluence of Ridley Brook (just short of 1 mile downstream from Bolton Falls Dam), the Winooski passes under a long railroad bridge. Fly fishers target the stream runs primarily downstream from the railroad crossing, since it's some of the nicest water in the area and is generally full of stocked rainbows and browns. You'll find a few road shoulder parking turnouts along River Road.

The railroad crossing reach during higher water levels. Located downstream from Bolton Falls, the reach is one of the most popular fly-fishing stretches. Located off River Road, other good river sections also flow through areas with good structure.

Continuing downriver, River Road becomes Duxbury Road and then Cochran Road. Cochran Road crosses a bridge that leads into Jonesville. The mouth of Huntington River, another important trout tributary (discussed in detail in later sections) flows into the Winooski a couple hundred feet downstream from the bridge. The Winooski flows nearly 7 miles downstream from Bolton Falls by the time it's joined by the Huntington River. Canoers and others with small floating craft can launch into the Winooski on the south side of the bridge that leads into Jonesville. The land is part of the Richmond Land Trust Nature Preserve.

Before reaching the Duxbury/Cochran Road intersection you'll pass a couple of access areas, some better for launching canoes or kayaks. Parking for Long Trail River Access (part of Camel's Hump State Park, just upstream from the mouth of Preston Brook) is located about 6 miles before the Cochran Road crossing. The large parking lot provides river access to anglers and other recreational users. A footpath leads from an informational kiosk down a short hill to the water's edge.

A couple other access areas are located downstream along Cochran Road. At the Duxbury Road/Cochran Road intersection near the bridge crossing into Jonesville, take a left. You'll pass a small bridge that crosses the Huntington River (a small parking area is located next to the bridge). Continue on Cochran along the south side of the Winooski. You'll pass Dugway road on your left (important for accessing the Huntington River—see page 93) before reaching the Warren and Ruth Beeken Rivershore Preserve and Overocker Park River Access areas. Both are located a couple of miles downriver from the mouth

The Winooski River relies heavily on stocked rainbow trout through much of its mileage. The fish are not particularly fussy about what fly pattern they'll strike. Streamers work well.

of Huntington River before the Winooski flows under another bridge crossing that leads into Richmond.

The river reach that leads downstream into the Richmond-Jericho area relies heavily on hatchery trout and is better fished out of a small floating craft. This is not prime trout water, but VF&WD continues its brown trout stockings that far downriver. About 1,800 browns are stocked annually from the mouth of Preston Brook downstream to the Bridge Street crossing into Richmond. Johnnie Brook Road under I-89 downstream to North Williston Road in the Richmond-Jericho reach receives about 1,000 brown trout annually. The river here is big water, where conventional tackle anglers target both trout and warmwater species.

Fly fishers cast a variety of fly patterns along the runs and pocketwater downstream from the railroad bridge crossing. The rainbows aren't particularly fussy about what flies are drifted in front of their noses, but olive or black Woolly Bugger patterns can do the trick. I use a 9-foot, 4-weight rod when fishing that reach during mid-spring. High-floating dry flies on-point with a Prince Nymph dropper can be effective. You'll run into caddis and mayflies on the Winooski, but nothing like the enormous hatches you'll encounter on rivers such as the Batten Kill.

LANDLOCKED SALMON

As it pushes into the Burlington area, the lowermost Winooski offers warmwater species experiences. Yet during the fall months lots of chatter surrounds the annual landlocked salmon spawning migrations from Lake Champlain. Lots of attention focuses on the Winooski One Hydroelectric Project. A landlocked salmon fish lift, in operation for more than 30 years, is operated between September and November a few times a day at the impassible dam. The salmon are put into tanks at the top of the lift, where biologists collect data.

Winooski One Dam in Burlington. Landlocked salmon swim up through the Salmon Hole area and into a fish ladder on the left side of the dam.

The fish are checked for lamprey scars, sex, length, and weight. Numbered tags are applied to the fish for population dynamics studies. (The 2023 numbers came in at a record high: more than 200 fish.) The landlocked salmon are then transported upstream by truck to the Winooski-Richmond area. Biologists attribute the increase in numbers to ongoing sea lamprey control in Lake Champlain.

You'll have plenty of company downstream from the hydroelectric dam during September into November at the Salmon Hole. You can access the Salmon Hole from 6-acre Salmon Hole Park, located off Riverside Avenue in Burlington. A small parking lot is located near the bridge to the city of Winooski. The more popular fly-fishing section is located on the Winooski side of the river.

Fly fishers routinely sample the Winooski downstream from the dam at other times of the year as well. Lots of other species worthy of targeting wander up in the area from Lake Champlain. While the Winooski doesn't attract a massive number of steelhead from Lake Champlain, fly anglers target that species in late winter. You might latch into anything while casting at the Salmon Hole—from landlocked salmon to steelhead to the occasional brown trout and other species. Take note of an important Winooski River fishing regulation at the Salmon Hole: The Winooski River is closed to fishing March 16–May 31 from One Hydro Dam west of US 7 in Winooski and Burlington, extending downstream to the downstream side of the first railroad bridge (to protect spawning walleyes).

Burlington resident and fishing guide Ralph Kucharek, proprietor of The Vermont Fly Fishing Company, has a good understanding of landlocked salmon fishing in Salmon Hole. But he'll be quick to tell you he prefers not to fish for landlocks along at the popular location. "We're spending a lot of money and effort to get natural landlocked salmon reproduction. It's probably better to leave them alone. Anytime from September on can be good for anglers who would like an opportunity to get into a good fish.

"My own experience there, with my own angling, is that you'll either blank out or hit it really big," Ralph said. "It's not unusual to latch into one in bright sunshine, and it's not uncommon for an angler to latch into an 8-pounder—even larger. I've also netted 6- to 7-pound steelhead in there. Soft hackle Hare's Ears, October Caddis-like flies work well. Salmon Hole is easily accessible. You can fish it from either the Burlington or Winooski side of the river. It's definitely one of the earlier runs of landlocks too. If landlocks are around, you're generally going to know. You might see them boiling on the surface or making a disturbance. Anglers who might not know exactly where to fish should look for any sign of activity."

Huntington River

Flowing south to north 18 miles from its headwaters in Buels Gore to the Winooski River, the Huntington River is one of six major tributaries that feed the Winooski. Its mouth at the Winooski is located just a couple hundred feet downstream from the Cochran Road bridge that leads into Jonesville.

If you're approaching from the mouth of the Huntington River, you'll want to get on Dugway Road. At the Duxbury Road/Cochran Road intersection near the Winooski bridge crossing into Jonesville, take a left. You'll pass a small bridge that crosses the Huntington River just upstream from its mouth (there's a small parking area next to the bridge). Continue on Cochran a short distance to Dugway Road, on the left.

Dugway Road is a dirt/gravel road that follows the Huntington River upstream a couple of miles until pavement resumes once it intersects Huntington and Main Roads. You'll pass the

Huntington River along the Green Mountain Audubon Horseshoe River Trail. Access to the stream is quick from the parking area in Huntington.

scenic if infamous Huntington River Gorge, a narrow chasm known more for the many lives it has taken than its scenic beauty. A parking area with an informational kiosk is located on Dugway, 1.7 miles from the Dugway Road/Cochran Road intersection.

A popular and well-known but dangerous swimming hole, signage is posted along the stretch to warn visitors of its dangers. More than 20 persons have drowned in its deceptive waters that lure recreational water lovers and others. Signage warns visitors to refrain from wading or swimming above the gorge, as slippery rocks can lead to potential disasters. Fly fishers should exercise extreme caution, even when sampling stretches well upstream from the falls. As you continue driving along Dugway Road upstream from the gorge, you'll pass some very nice water, much of it located down a short yet steep hill. There's easier stream access upstream.

Take note of an important informational kiosk located at a small parking lot along the gorge area.

A short footpath leads down a manageable slope to the beginning of the gorge. Admire the plunge pool, but locate elsewhere upstream for safer fishing.

GREEN MOUNTAIN AUDUBON RIVER TRAIL ACCESS

Dugway Road intersects Huntington Road about 1.6 miles upstream from the gorge parking area. Take a left onto Huntington Road and you'll shortly get to nice fly-fishing water. In about 0.5 mile Huntington Road becomes Main Road as you enter Huntington. A short distance after passing into Huntington, just after a bridge crossing, you'll find the parking lot for the Green Mountain Audubon Sugarhouse. A trail across the road follows the river close to 0.5 mile, ending at the Green Mountain Audubon Horseshoe parking lot.

The runs and riffles along the River Trail provide very pleasant fly fishing for hatchery rainbows that are stocked in sections upstream from the gorge into Huntington. Brook trout are stocked from the bridge crossing just north of Huntington Center upstream to Beane Road in Hanksville. Trout rise willingly to Elk Hair Caddisflies skittered across the runs. Cast small bead-head flies into the small pools.

Little River below Waterbury Reservoir

One of the most interesting Winooski River tributaries is Little River. Its popular lower reach, just west of Waterbury Village, is a short tributary, barely 3 miles in length as it flows as a cold tailrace from Waterbury Reservoir. An important spawning tributary for the Winooski River, it's managed as a wild trout fishery. Upper reaches above Waterbury Reservoir also support trout.

At Waterbury Reservoir, also a splendid fly-fishing water, Little River State Park is located up Little River Road a short distance north of Waterbury Dam. The park is a nice place to camp at one of its numerous RV/tent sites, rental cabins, and lean-to shelters. Waterbury Center State Park is also located on the reservoir. While visiting and fishing Little River, consider casting flies on Waterbury Reservoir. Brown and rainbow trout are

Little River regular and Burlington area resident Tom Marble particularly likes fishing the stream during winter and early spring, times he can have the water almost to himself.

stocked in the reservoir. Smallmouth bass fishing can be good. Of benefit to canoers and kayakers are the two 5-mph wake-free zones on the lake.

What makes this tributary most interesting, besides its wild browns and rainbows, is how it was transformed into a wonderful little trout stream following natural disasters that struck the area back in the late 1920s and mid-1930s. Little River State Park is situated on land that was once a thriving farm community. More than 100 years ago, some 50 families worked the land, sawed timber, built schools, and created businesses on the vast acreage. By the late 1800s, families in search of better farming soil and living conditions began abandoning their farmsteads. Then came the floods of 1927 and 1934, driving out almost everyone except a guy name Charlie Ricker. W. Patrick Yaeger's book *That Beautiful Vale above the Falls* tells the story of the community and Charlie Ricker's refusal to leave the area.

A couple of years after the floods, the US Army Corp of Engineers along with the Civil Conservation Corps (CCC) constructed Waterbury Dam. Today the once-thriving farming community is a ghost town. Visitors to Little River State Park can walk trails that pass ruins, relics, and even the old cemetery. Charlie Ricker's house still stands after 120 years, now on Vermont State Lands and maintained by volunteers. The coldwater tailrace discharged from the dam enters the Winooski River a couple of miles downstream, enhancing that fishery.

You can fish the lowermost section of the stream by taking Farr Road off US 2 before arriving at Little River Road. An angler parking lot is located near the I-89 overpass. VF&WD signage at the lot reminds anglers of the Special Regulations. After sampling the mouth area, take Little River Road to various stream areas that can be accessed at parking turnouts and lots. You'll encounter signs posted at first, but these quickly vanish as you make your way upstream toward the dam. Little River has a mix of nice runs and pools, although some stretches are easier to wade and fish than others.

The Anglers' Trail information kiosk is located at streamside just before reaching Waterbury Reservoir Dam on River Road.

JC's Electric Caddis *Tied by John Collins*

John created this pattern after browsing Ted Leeson and Jim Schollmeyer's *The Fly Tier's Benchside Reference* (1999). He tied it in various sizes on a Daiichi hook model 1220. The abdominal claw is created with green fluorescent Antron yarn. Chartreuse Ultra Wire inserted into hollow Wapsi Stretch Tubing forms the body. Brown ostrich herl forms the legs. The pronotum is herl burned on top then coated with Solarex UV resin.

Start at the dam area and work your way downstream. Just before reaching the dam, where a gate is located, a parking spot next to the Anglers' Trail information kiosk is convenient to a short footpath that leads to the stream. That's where I met Burlington area resident and enthusiastic Little River fly fisher Tom Marble.

"I fish this river all winter," Tom told me on a sunny day in late March a couple of years ago. "I've even driven up here at times before Little River Road had been plowed after a snowfall." One advantage of fishing Little River during what other fly fishers consider off-season is avoiding the oft-crowded stream during late spring and summer. The fishing isn't easy, targeting the wild browns and rainbows that swim in its waters. During the depths of winter, Tom often casts streamers through the deep pools. "The trout are probably focused on baitfish that gather in deep water," he explained.

During that day in March, Tom was focused on fishing caddisfly larvae imitations. He opened a fly box that was filled with various caddis-type flies that he fishes deep along the runs. He'll also fish a small size 22 midge fly on a dropper and sometimes a Walt's Worm pattern on-point. He's found that Little River is predominantly a caddis/midge stream. Tom also fishes Electric Caddis larvae flies.

Little River—West Branch above Waterbury Reservoir

While lots of attention is directed to Little River's short mileage downstream from Waterbury Dam to the Winooski River, for good reason, the West Branch's upper reaches above the reservoir and important tributaries also are worth exploring. Not far from the hubbub of Stowe, Vermont, arising from Adams Apple on Mount Mansfield, the West Branch of Little River begins its decent from its 3,200-foot elevation. At 4,393-foot elevation, Mount Mansfield is the highest mountain in Vermont, recognized for its true alpine tundra environment. There's some confusion surrounding exactly what the stream's name is. Locals refer to it simply as "West Branch"; others stick with "West Branch of Little River." However, official USGS topological maps show "West Branch Waterbury River." The latter could have been residual nomenclature that goes back before Waterbury Reservoir was constructed.

Access is available by taking VT 108 (Mountain Road) from Stowe and following the river upstream. Crossroads, parks, and other areas will get you into the water. One nice stretch is available at the Brook Road crossing, downstream from the covered bridge. It's located about 4.3 miles up VT 108 from the VT 108/VT 100 intersection in Stowe. A small parking area is located next to the covered bridge. During the busy tourist months, the lot is usually jam-packed with vehicles, and lots of mountain bikers park at the lot, a launch to the Stowe Recreation Path. It's probably best to avoid the area during summer unless you can get on the water very early in the morning. Crowds can be incredible along the entire VT 108 corridor during summer.

West Branch of Little River at Brook Road crossing. Covered bridges span many streams throughout Vermont, including the West Branch of Little River, not far from Stowe.

Mountain bikers, trail hikers, and other recreational users target the area and its many side attractions, including various inns and resorts. VT 108 (Mountain Road) leads to one of the most visited recreational areas in Vermont—Smugglers' Notch, a mountain pass. Mountain Road winds and turns its way upward and through Mount Mansfield State Forest.

By the time the West Branch joins the Little River at Stowe (USGS topological maps label it "Waterbury River") after roughly following VT 108 downstream, it will have dropped 2,500 feet. Along its southeasterly path it picks up Ranch Brook, a splendid little brook trout water. In Stowe, Little River makes a turn and briefly flows along VT 100, accepting Gold Brook just south of Stowe before taking another brief turn northwesterly on its final approach to Waterbury Reservoir.

Gold Brook also harbors wild trout along with a mystery from the past. The famous Gold Brook Covered Bridge, better known by some as Emily's Bridge, was once featured in the paranormal TV series *Most Terrifying Places in America*. The story goes back to the 1850s, when a distraught bride-to-be supposedly killed herself by jumping from the bridge into the brook. You're not likely to encounter her ghost, as local folklore tells us, but you could very well encounter a nice wild brook trout below the bridge. An interesting tidbit about Gold Brook has to do with something other than trout. The last time I was down in the area near Emily's Bridge, I spotted a lady down in the stream panning for gold.

The covered bridge is located a few minutes south of Stowe, where Gold Brook and Stowe Hollow Roads meet. You can park at the bridge in a small turnout on the Stowe Hollow Road side. Good fishing can be had along its downstream stretches, not far from the VT 100 crossing.

Emily's Bridge over Gold Brook is the site of local haunting folklore. Gold Brook holds wild trout.

Pocketwater directly beneath the Gold Brook Covered Bridge, where folklore says Emily took her life.

VF&WD has identified a number of upper Little River watershed tributaries that are important spawning and nursery waters. The following waters were listed in the *2017 Basin 8—Winooski River Watershed Water Quality and Aquatic Habitat Assessment Report*, prepared by the Vermont Agency of Natural Resources and Department of Environmental Conservation. The West Branch of Little River supports wild brook trout above the confluence with Ranch Brook and wild brook trout and brown trout below. Other important small tributaries include:

- **Alder Brook:** Joins Waterbury Reservoir's east arm—wild brook trout
- **Stevenson Brook:** Joins the southwest end of Waterbury Reservoir—wild brook trout, brown and rainbow trout; rainbow smelt spawn within the confluence to Waterbury Reservoir
- **Miller Brook:** Joins Little River just upstream from Waterbury Reservoir—wild brook trout, brown and rainbow trout
- **Gold Brook:** Joins Little River just south of Stowe—wild brook trout, brown and rainbow trout
- **Ranch Brook:** Joins the West Branch northwest of Stowe—wild brook trout

Ranch Brook

Born in Mount Mansfield State Forest, home to its namesake mountain—at nearly 4,400 feet, Vermont's highest—Ranch Brook ends where it weds the West Branch of Little River, not far away. Their combined water eventually contributes to the Winooski River. Access the stream by taking VT 108 north to Stowe Fork; Ranch Brook Road will be on your left. Then take an immediate right onto the rocky dirt road that follows the brook.

Brook trout that inhabit Ranch Brook are not large, but they are willing gems that will strike Elk Hair Caddis as well as other small dry flies and nymphs. Most of these wild brook trout are french-fry size, but dollar bill–size fish exist. Keep in mind that the brook runs through a multiuse recreational area close to one of the most visited areas in Vermont. Running though the 513-acre Adams Camp property, owned by Trapp Family Lodge and conserved by Stowe Land Trust, this splendid brook trout stream is the perfect water for your 6-foot, 3-weight rod.

Ranch Brook, in the Stowe area, is a lovely little stream perfect for hip boots or summer wet-wading. Small yet willing wild brook trout inhabit its small pools and runs. VALERIE VALLA

Stowe has been called a "premier destination for outdoor enthusiasts," not only during winter, when skiers flock to the "Ski Capital of the Northeast." The area is also busy during summer. During summer's heat you'll encounter trail hikers dipping into Ranch Brook's cool, clear waters and dog walkers taking early-morning treks up Ranch Brook Road, the dirt and oft-rocky road that follows the brook partway to a summer parking spot near a trail footbridge. Despite the area's popularity, I most enjoy sampling the brook during late summer into early fall.

Anglers and other recreational users aren't the only ones interested in the stream. VF&WD has been monitoring Ranch Brook since the 1990s. Since it's a trout water that isn't impacted by development along its length, the stream provides biotic insights on how the stream system may have changed over time compared to other streams that don't flow in a similar more-natural environment. VF&WD fisheries biologist Bret Ladago provided a good overview of what's being monitored, including water quality and temperature that's monitored using logging devices put in the stream every year. Temperatures are recorded by the hour, every day. Electrofishing is also done to monitor brook trout populations over time.

Mad River

If approaching from the Champlain Valley and New Haven River (see page 82), you'll most likely choose Lincoln Gap Road out of the Bristol-Lincoln area to get you over the mountain to the Mad River Valley. At 2,424 feet, seasonal Lincoln Gap Road is the highest mountain pass accessible by vehicles in Vermont. (Lincoln Gap Road is typically closed and barricaded from mid-October to about mid-May.) From Lincoln and VT 116, turn onto Lincoln Gap Road. A very steep climb toward the pass, an almost 25 percent grade, the road will eventually begin its downhill, equally snaking path, meeting VT 100 in Warren,

A short stretch of Mad River located off Old Route 100 is accentuated with boulders and rocks that provide good habitat structure for rainbows.

headquarters for fishing Mad River. If approaching from other areas near Montpelier and other parts of Vermont, other routes deliver you to Mad River, but you'll want to get to VT 100B/VT 100, the Mad River Scenic Byway.

Born at Granville Notch in the Granville area, the roughly 26-mile-long river primarily follows VT 100B/VT 100 on its south-to-north path, passing through the villages of Warren, Waitsfield, Moretown, and finally Middlesex, where it joins the Winooski River not far from Walter H. Kelley Park. Along its path, anglers share Mad River with recreational canoers, swimmers, and others drawn to the scenic Mad River Valley. During winter you'll encounter plenty of vehicles lugging ski equipment on their roof racks.

Despite the popularity of the river with other recreational users, there's plenty of water available for fly fishers to enjoy good sport. Mad River is predominately a stocked rainbow trout stream, although wild brook trout are found in its headwaters and brown trout inhabit the lower river pools. Several small tributaries that enter the river along its course also harbor fish.

GRANVILLE AREA TO UPSTREAM FROM WARREN VILLAGE

In Granville, on VT 100 where it's joined by West Hill Road, Post Office Hill Road, and Mill Road, you'll notice a small stream running along VT 100. That's the White River headwaters that, like Mad River, is also fed by waters off Granville Notch. White River heads south along VT 100 through the Connecticut River Basin (see page 170). Mad River, once it develops and shows itself close by on VT 100, flows north, the opposite direction.

If you're approaching Mad River Valley from the Granville area, continue north on VT 100 from the road intersections. At first VT 100 follows Alder Meadow Brook, a White River tributary, for a few miles before Mad River begins in a brushy area at Granville Notch. In about 6 miles you can sample the headwaters by parking at a small road shoulder turnout. Walk down the road a short distance for easier access into the stream. In about 7 miles or so, at the Addison County–Washington county line bridge crossing, you'll begin to get a better idea of Mad River's growing appearance. Parking and stream access is easy at both the bridge crossing and a small streamside picnic area just after the bridge crossing. Like many other streams in Vermont, Tropical Storm Irene tore through Mad River, including this headwater area. Bank stabilization rocks have been placed here along the still small stream.

In about a mile past the streamside parking area, Stetson Brook joins the river, adding volume to Mad River as it continues through a forested narrow section that provides good habitat for wild brook trout. Just over a mile past Stetson Brook on VT 100, the Warren Falls trailhead is on the left. (If you're approaching from Warren downstream, the trailhead is located about 1 mile south on VT 100 from the Lincoln Gap Road intersection.) Picturesque Warren Falls becomes a zoo during summer, with swimmers and other recreational users occupying the area and numerous vehicles parked along both sides of the road. If you want to fish that section, summer is not a good time. The same holds true for many areas along Mad River.

Fly angler Colin Phillips, who has fished Mad River since boyhood, told me he has netted brown trout downstream from Warren Falls to Bobbin Mill. He has also explored areas upstream from the falls. I had the pleasure of renting his home along Mad River during early May and was able to learn a few things about the Mad River system I wasn't aware of.

"I've only caught browns below the falls in a stretch just below the falls that extends down Bobbin Mill. I'll usually fish it from Bobbin Mill up. Above the falls I've only ever caught brook trout. It can occasionally be decent up that way, depending on whether or not the beavers have done any work and created some nice pockets. If you go further up from there, there are some remote camping sites on the right in the Green Mountain National Forest, and you'll find a few tiny streams, places I grew up fishing with my dad, learning as a kid. Stetson brook can be productive at times, though, like most areas, it got nailed by Irene. But if you can get in there before it is wormed, it can be fun for native brook trout."

I had been communicating with VF&WD biologist Bret Ladago for some time concerning various tributary waters, such as the Dog River. Bret knew that I had been actively performing DNA sampling in the area and asked if I was planning to sample Mad River. Apparently, an angler had reported the presence of rainbow trout above Warren Falls, an unusual finding. I was staying near Warren Falls for a few days that week, so I gladly ran DNA sampling at a site above Warren Falls at coordinates Bret asked for the sample to be taken. While I've said all along that DNA results should always be examined with caution when it comes to certainty, *Oncorhynchus mykiss* (rainbow trout) was detected. Not very much, but data showed a 100 percent base-pair DNA match. Of course, lots of brook trout DNA also was detected.

WARREN VILLAGE TO WAITSFIELD COVERED BRIDGE

A short distance downstream from Bobbin Mill and the Lincoln Gap Road intersection with VT 100, you'll enter Warren Village, where Mad River continues its traverse through areas that change from gorge-like settings to calmer waters. A drive down Covered Bridge Road off VT 100 will take you over a quaint covered bridge built in the late 1800s and onto Main Street. Mad River cascades over a dramatic old mill dam site waterfall along Main Street. Remnants of the dilapidated dam are in view, torn apart by floods over the years.

After crashing over the dam and through a gorge-like setting, which can be quite dramatic after heavy rains, the stream then flows through a calmer stretch that can be very good fly-fishing water. Park on Main Street next to the Park and Ride lot, but you'll want to access the stream on the far side. You can get into the stream by walking down the street a very short distance to the bridge crossing. Walk across the bridge to Trout Hollow Road and get into the water. It's fishable upstream to the falls, depending, of course, on water levels. Swimmers and young waders get into the water near the falls during lower-water summer conditions. When you walk or drive across the bridge, you'll surely notice Mad River continuing its brief crash downstream, where yet more interesting fly-fishing stretches await.

In less than a couple of miles downstream from Warren off VT 100, the Mad River flows along the Quayle Conservation Area. A nice run enters a long, flat pool that's popular with other recreational users. Just downstream from the Quayle Conservation Area, you'll reach the Kingsbury Greenway—Wabanaki Conservation Area entrance off VT 100. A good fly-fishing strategy is to park at the large conservation area lot near the kiosk at its entrance. You'll notice a narrow stone footpath that leads to the bridge crossing. The path actually continues as a formal trail that passes beneath the bridge and continues upstream to Riverside Park.

The water along this reach is conducive to fishing size 12 hair wing dry flies that can be worked along the riffles while making your way upstream to Riverside Park. You can then reverse course and fish back downstream to the conservation area.

While wild brown trout inhabit some sections of Mad River downstream from Warren, many anglers target the rainbow trout that are stocked in the river. VF&WD typically stocks 1,300 rainbows that approach 12 inches from the Warren Village area downstream to the Waitsfield covered bridge, about 6 miles downriver.

The Quayle Conservation Area reach is located off VT 100, a short distance downstream from Warren. It is best fished during May, before other summer recreational users also enjoy the scenic stream reach that spills into a nice pool.

WAITSFIELD COVERED BRIDGE TO VT 100B BELOW MORETOWN

As you approach the Waitsfield area, still driving north from the conservation area, sample the run along Mill Brook that you'll notice just before reaching the VT 100/VT 17 intersection. Mill Brook is an important spawning tributary. You can park along the stream at a small shoulder turnout. It's a nice stretch that often harbors rainbows that hold along the riffles and runs. Once you get downstream from Waitsfield, the river tends to open up and get warm by summer, something Colin mentioned. "It gets pretty flat in the Moretown area and opens up quite a bit, but they stock the heck out of it regardless." However, try the Moretown reach at the falls (you can view the falls from the Fletcher Road bridge crossing). Colin recommends getting into the river downstream from Fletcher Road. "There is a sandpit almost directly across the Moretown Road Department building in Moretown. You can park in the back and hike up along the river edge to Moretown Falls. The stocked rainbows tend to congregate just under those rocks in the spring." Another stream reach with decent rainbow holding water is located about 1 mile downstream from VT 100B/Fletcher Road, just beyond Sugarhouse Way. A short pullover along the road shoulder can accommodate anglers, but because of its easy access to a nice section of the river, expect company. Another small road shoulder parking area is located less than a mile downstream next to Kingdom Farm Lane.

Interesting water also flows along Old Route 100, which briefly departs then rejoins VT 100B less than a mile downstream from the Kingdom Farm Lane intersection. Keep watch for posted signs along this reach. A half mile downstream from Old Route 100/VT 100B, the wide-open river runs through a popular recreation area. "You'll hit Ward's Park, or Ward's Swimming Hole, as we call it. There are some massive boulders there you can see from the road. I've caught many trout there on the fly and spinning reel as well. This spot used to have holdovers, but I think those days may be over. Nonetheless, after stocking, it's a hot spot," Colin said.

The parking area at Ward's leads to a long set of stairs that end at the riverside swimming beach area. You'll want to stay clear of this area during summer, for certain. In less than a mile north of Ward's, VT 100B will get you to the Moretown No. 8 Hydro Dam near the river's confluence with the Winooski River at Middlesex. While the dam is quite far downstream, the stretch below it can hold trout.

Fly angler Clark Amadon, who has resided in the Mad River Valley for 26 years, shared this about the Mad River: "The convergence of the Mad and the Winooski in Middlesex offers excellent fishing. Here an angler can swing wet flies, float an indicator rig, rip a streamer, and skate caddis patterns. *Isonychia* mayflies are a great fly to imitate almost all summer long. One can use a 'Iso' dry fly pattern, a prince nymph or a Leadwing Coachman in this reach. Fishing a prince down and across is very effective here! In the fall, large browns can be had up from the confluence with the Winooski to Lovers Lane Bridge. It's flat and clear water, so wade with care and look for subtle rises."

Clark, a former chair of the Vermont Council of Trout Unlimited who has held all TU officer positions aside from treasurer, is a passionate conservationist who holds the Mad and other area streams near to his heart. Besides fishing the lowermost reaches at Mad River's confluence with the Winooski, he enjoys the river's headwaters as well. "As is usual for headwater streams and tributaries, the trout respond to largish high-floating dry flies. On hot summer days, escaping to the upper regions is delightful! Wet-wading for wild trout and cooling off in the process," Clark said.

Dog River

Considered one of the best wild trout streams in all of Vermont, Dog River (one of Vermont's Special Regulations trout waters) supports a population of rainbow trout, brook trout, and sizable brown trout. Its upper reaches support wild brook trout. Situated in the upper section of the Winooski watershed, "The Dog" begins its journey near the town of Roxbury at 2,300 feet in elevation. The stream then flows north, traversing through a gentle gradient valley with bedrock gorges along the way, dropping to about 500 feet in elevation where it weds the Winooski River in Montpelier, Vermont's capital.

Along its approximately 19 miles, the pleasant stream picks up water from a number of small brooks, all managed as wild trout waters—Felchner, Sunny, Stony, Union, and Cox among them. Bull Run, a small tributary that empties into the upper Dog River along VT 12A, has a particularly interesting story. The Dog follows VT 12A in its headwaters, then continues flowing along VT 12 all the way to Montpelier. Along its course it flows through Northfield (the home of Norwich University), Northfield Falls, and past the Berlin-Riverton area on its way to the Winooski River.

HEADWATER REACHES ALONG VT 12A

One way to approach fishing the Dog River is to locate in Northfield and then first sample the stream's headwaters, fishing along its riffles and runs. Turn onto VT 12A at its intersection with VT 12 in Northfield. The road follows the river upstream 6 miles or so to near its source in the town of Roxbury. Not all the headwater stretches along VT 12A are conducive to fly fishing.

At about a mile's drive south on VT 12A, a small shoulder parking spot next to the Fairground Road bridge crossing can get you into the straightened river. It's not been a

Dog River at the Water Street reach in Northfield. Dog River Park, conceived and developed after catastrophic flooding from Tropical Storm Irene inundated the area in 2011, is situated a couple hundred feet downstream from this section. Streambank stabilization has been placed in the area.

favorite reach of mine, but it's worthy of a couple of casts along its riffle structure. Bull Run tributary enters the stream about 0.5 mile or so upstream; you'll see it entering the Dog just downstream from the overhead railroad trestle.

Removing the old defunct dam on Bull Run is one of the best examples of a conservation/trout habitat improvement project and is helping Dog River maintain a great wild trout fishery. It took three years and $400,000 to get rid of the 100-year-old dam, described as a crumbling mess, that prevented trout passage into its cool thermal-refuge waters.

The bridge was finally removed in 2020, with Friends of the Winooski River spearheading the project. Funding and technical assistance came from multiple partners, including Eastern Brook Trout Joint Venture, Lake Champlain Basin Program, The Nature Conservancy, US Fish and Wildlife Service, Vermont Department of Environmental Conservation, and Vermont Natural Resources Council.

Another access point is located just 0.1 mile upstream from the overhead railroad trestle. There's a small parking turnout on the left side of VT 12A, just across from Stony Brook Road. If you're going to park at that location, be mindful that others frequently pull over here to fill their bottles with the natural spring water that runs out of a conduit from the hillside. Please don't block others who want to access the spring. Driving another mile south on VT 12A will get you to another small shoulder parking turnout on the left side of the road. Fish the water along that stretch with small bead-head flies. You'll run into increasingly small water along the remaining 3.5 miles upstream to Roxbury, not conducive to fly fishing.

NORTHFIELD REACHES AND ACCESS ALONG VT 12

Try Dog's various stretches around Northfield itself, as well as water that flows in an area behind Norwich University, by heading back to the VT 12A/VT 12 intersection. From the intersection, drive about a mile north on VT 12, past the university, to a bridge crossing next to a Dollar General store. Take a left on Water Street and continue to Dog River Park. You'll see a small pavilion on the left next to a large open field maintained as a pleasant natural area. Park on the roadside and walk down one of the paths (used more by dog walkers than anglers) to the stream. When Tropical Storm Irene ripped through Dog River in 2011, what is now an open field once was lined with homes that were severely damaged by the catastrophic inundation.

This area was always prone to flooding, but Irene really set a change in motion. FEMA assisted dwelling owners with buyouts of structures that were too damaged to repair and restore. What is left today is easy river access for anglers and a nice environment for folks who now enjoy a pleasant walking area next to the stream. It takes a bit of a hike, but you can try fishing areas upstream from the park or head downstream and fish through the area that leads back to VT 12. The drawback is the congestion of buildings along the way, not something I favor myself. I took a DNA sample in the area of the small pavilion. An abundance of brook trout, black nose dace, long nose dace, and a very small amount of brown trout DNA was detected flowing through the reach. The prevailing fish species detected were members of the Cottidae family—sculpins. Better water is found downstream from Northfield.

A couple of nice river reaches with easy access are located along VT 12. Try the stream reach behind the Dollar General store, located down a hillside at the breached old Cross Brothers Dam. The dam has received considerable attention in recent years. The Vermont Natural Resources Council received funding for its removal from the National Fish Habitat Partnership. US Fish and Wildlife Services and others also are involved. Coldwater brook trout habitat will be greatly enhanced, benefiting the entire system. You'll notice a large pool at the breached dam's spillway as it now flows (2024). Bret Ladago, a VF&WD fisheries staffer, told me recently that rainbows have not been seen above Cross Brothers Dam since 2006.

If the Cross Brothers Dam stretch doesn't prove productive, continue north on VT 12 to Dog River Drive, less than a mile from the VT 12/Water Street intersection. Park next to the bridge crossing across from the Town Highway Department building. Fish downstream along the gentle grade riffles with small flies or upstream from the bridge crossing. I took a DNA water sample at that location in mid-February 2024. Results showed brown and brook trout DNA flowing through that reach, but *Oncorhynchus mykiss* (rainbow trout) was not detected. Macroinvertebrate kick samples revealed a healthy assemblage of stoneflies and mayflies, particularly *Ephemerella subvaria* (Hendrickson/Red Quill) nymphs.

Red Quill DOA Cripple Emerger *Tied by Henry Ramsay*

A nice stretch on middle Dog is off Cox Bridge Road in Northfield Falls, less than a mile north of Dog River Drive. You'll want to get your lunch sandwich at Northfield Falls General Store at the corner of VT 12 and Cox Bridge Road. Parking alongside the store is restricted to patrons. The challenge

A nice stretch of Dog River flows along an area about 0.5 mile north of the Cox Bridge Road crossing.

to fishing the beautiful area along the pleasant narrowly confined, steep bank valley stretch that gathers water from Cox Brook—accentuated by quaint red-painted covered bridges—is parking. One very small Cox Bridge Road shoulder parking spot is located just beyond the first covered bridge and the end of a guardrail. You'll no doubt be competing with other recreational users, especially in summer, for parking there. Arrive early in the morning during May and again during the early fall season, when there's less competition. Wading can be a challenge during spring fishing but easier in late season, when water levels are lower.

About 0.5 mile north of Cox Bridge Road, on VT 12, park along the wide road shoulder for access into some nice stretches. Another couple of miles north on VT 12 will take you to a green bridge crossing. Just across the bridge, next to the Berlin Fire Department building, a path will take you down the hill to a swimming hole popular with summer users. A bench overlooks the river, and there's a picnic table where you can eat the sandwich you purchased at Northfield Falls General Store. You'll notice VF&WD signage at the footpath that reminds anglers of Dog River's Special Regulations status.

About a mile beyond the Berlin Fire Department, still on VT 12, you'll come across a small shoulder turnout that can get you access into the river. (Make sure you don't block the water-intake hydrant at the small parking spot.) Fishing can be good along that stretch. Other access areas are available downstream from that reach, some at bridge crossings as you continue north on VT 12. Less than 4 miles north on VT 12 from the small parking spot, turn left on Dog River Road (not to be confused with Dog River Lane, upstream in Northfield). A short drive will get you to a river bridge crossing on Nelson Road. When you get that far downriver, very close to Montpelier, you'll note the very substantial slow, deepwater holes that undoubtedly harbor big trout. Try the run accessed on the far side of the bridge with large streamers. Dog River Recreation Area soon comes into view, not far from the Dog River's confluence with the Winooski at Junction Road. The DNA water sample I took downstream from the Nelsen Road crossing in April 2024 revealed a good amount of rainbow trout DNA along with something I didn't expect. In the mix with various other fish species, including brook stickleback (*Culaea inconstans*), was *Sus scrofa*—wild boar—DNA. Wild boar have been sighted roaming parts of Vermont.

Lamoille River

The Lamoille River begins its life at Horse Pond, a shallow 32-acre warmwater fishery, then flows 85 miles to Lake Champlain, where it dumps into Malletts Bay north of Burlington. As part of its Vermont River Watershed *Basin 7 Tactical Basin Plan*, the Vermont Agency of Natural Resources conducted water temperature surveys along the Lamoille River and 11 tributaries between late May and mid-November a few years ago to better understand habitat available for wild trout populations.

Many of the tributaries, including small brooks, were also surveyed to ascertain natural trout population numbers. Wild trout predominance, brook and juvenile rainbow trout, varied greatly in the tributaries, but one small brook that feeds Lamoille's headwaters showed a very large number of individuals, hitting a few thousand per mile.

While wild trout are netted frequently along Lamoille's mileage, in general terms the reaches from Peterson Dam in Milton downstream to the river's confluence at Mallett's Bay are considered a warmwater fishery between June 1 and September 30. Upstream reaches are considered a coldwater habitat. Much of the river, even flowing along its headwaters, experiences elevated water temperatures during summer. The Lamoille River is heavily dependent on supplemental stockings of rainbow trout. There wouldn't be much of a trout fishery on the Lamoille without hatchery fish.

Willy Dietrich, owner of Catamount Fishing Adventures in Stowe, Vermont, is one of Lamoille River's most knowledgeable fly fishers and guides. His nearly 30 years' experience

The Lamoille River along the East River Road reach in Johnson. This popular section of the river is particularly scenic and provides good fly fishing for rainbow trout. VALERIE VALLA

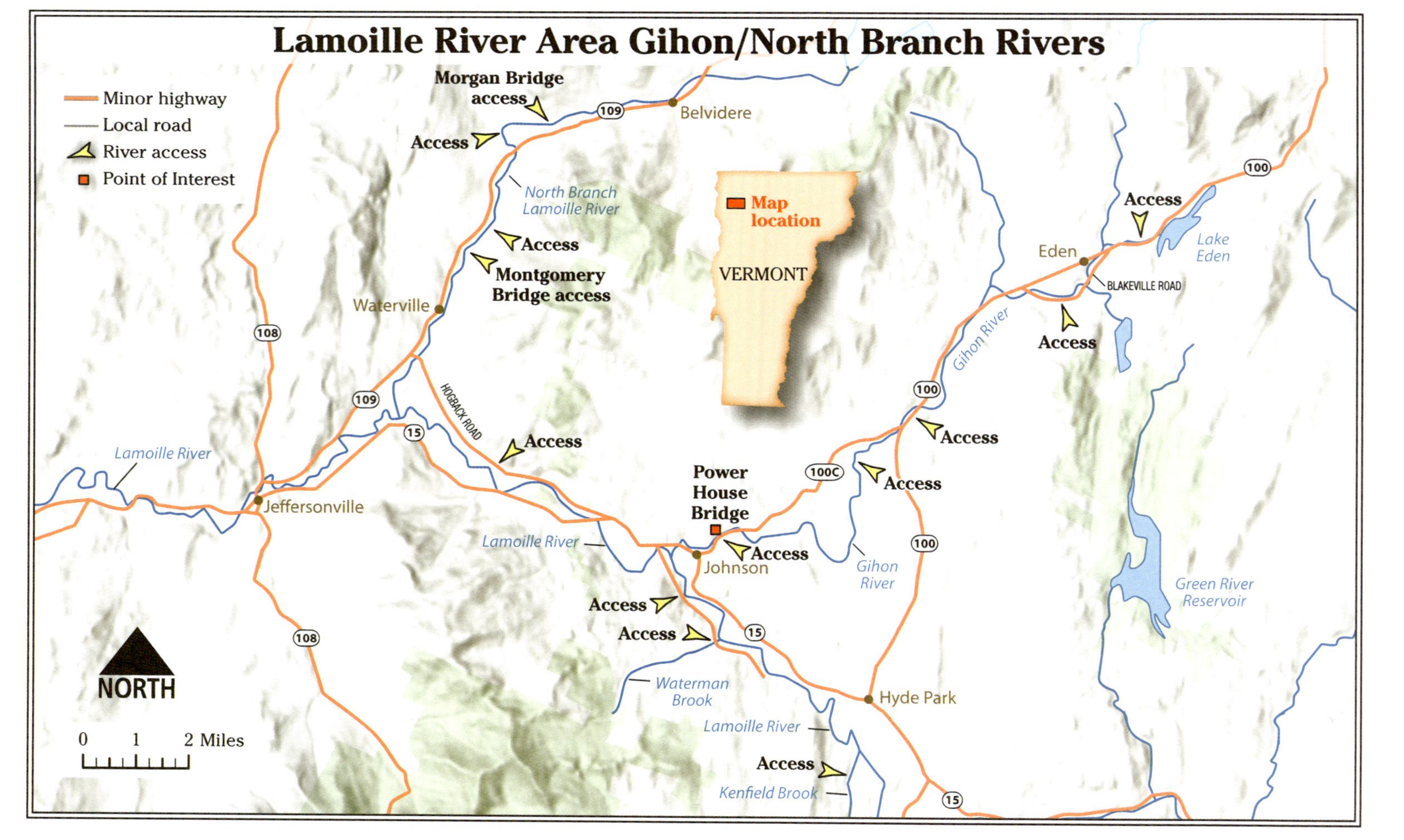
Lamoille River Area Gihon/North Branch Rivers
Minor highway
Local road
River access
Point of Interest
Morgan Bridge access
Access
109
Belvidere
North Branch Lamoille River
Access
Montgomery Bridge access
Waterville
108
109
15
HOGBACK ROAD
Access
Map location
VERMONT
Lamoille River
Jeffersonville
Lamoille River
Power House Bridge
Access
Johnson
Access
Access
15
Waterman Brook
Lamoille River
Access
Kenfield Brook
Hyde Park
15
108
NORTH
0 1 2 Miles
100C
Access
Access
100
Gihon River
100
Gihon River
Eden
BLAKEVILLE ROAD
Access
Access
100
Lake Eden
Green River Reservoir

The Lamoille River is born at Horse Pond, a very shallow 32-acre mostly perch and pickerel pond north of East Greensboro off VT 16.

guiding the river and other area waters provides an interesting perspective of how the river has changed over the years. "The Lamoille has real issues, and I consider it a very mediocre trout river," Willy told me recently. "It is largely a stocked trout fishery. The wild trout population in the main stem is in peril. The State of Vermont allows anglers to harvest eight trout. It is stocked with rainbows. There are wild brown trout, yet they seem fewer and far between. Hatches are okay on the Lamoille. Lots of caddis and a select few mayflies and stoneflies. Still, I do not witness the hatches I did even 20 years ago."

While Willy laments the changes that have occurred on the Lamoille River over the years, which he attributes to a number of factors, he still enjoys guiding on the stream, including reaches downriver for bass, pike, and landlocked salmon. He especially enjoys fishing high-elevation tributaries, where he targets brook trout. "My favorite venues are small high-elevation streams/brooks. The beauty of our small streams is they are nutrient poor. We grow rocks well in Vermont. Not a lot of matching the hatch. So lots of blind fishing and prospecting with big gaudy dry fly patterns or ripping heavy streamers into plunge pools." One important point Willy made was that he doesn't overfish the tribs. "I typically don't return to a spot I've fished for at least a week."

Willy's go-to dry fly for a small-stream trout is a size 10–12 rubber legged Royal Trude. "I fish lots of Wulff patterns, terrestrials, and silly attractors. Yet the Trude is always my go-to. These fish would eat pocket lint on a hook if they are not spooked. Pretty greedy. Of course, due to the lack of food and a short growing season, the trout tend to range from 4 inches up to the occasional 12- to 15-inch specimen. Perfect for short slow-action 2-weight and 3-weight rods."

UPPER REACHES FROM GREENSBORO TO WOLCOTT

After departing its birthplace in Horse Pond, the tiny Lamoille begins its long journey at first flowing southwesterly along VT 16, picking up a few small tributary brooks along the way. Among them, Page, Flagg, and Stannard Brooks contribute their small volumes. Caspian Lake, a splendid 800-acre trout water known for its natural populations of lake trout that thrive in depths that reach over 140 feet, is in the area. Fly anglers know Caspian Lake for its prolific *Hexagenia* hatches from late June to early July. However, natural brook trout and rainbows swim in its waters; it's no longer a stocked lake. It's a popular vacation lake, with a public launch where anglers, swimmers, and boaters enjoy the surrounding scenery. After passing the Greensboro Bend area and East Hartford, the river makes a turn and follows VT 15 in a westerly direction. A couple of headwater stretches along VT 15 are worth a look.

About 5 miles down VT 15 from Horse Pond, take a left onto Main Street and park next to the intersection, where you can get into the water. You'll want to use a shorter rod and cast small nymphs and small streamers into the runs. Stannard Brook, once one of the easier headwater tributaries in the area to fly fish for small brook trout, is located about 1.6 miles from the VT 15/Main Street intersection that leads into Greensboro Bend.

Located about 7 miles downstream from Horse Pond and about 4 miles northeast of the VT 16/VT 15 intersection, Stannard Brook took a brutal beating in July 2023 from a two-day rainstorm flood that devastated the area. Storms also struck the area in July 2024. The little stream now flows through a mess of jumbled trees that clog the runs along many uppermost stretches. However, its lower area at the Stannard Mountain/Orton Road intersection bridge crossing in Greensboro Bend is worth a few casts. Park at the road shoulder on Orton across the bridge. I performed a DNA sample at the bridge crossing, and the results showed a healthy amount of brook trout DNA flowing through.

Caspian Lake, a popular recreational lake near the headwaters of Lamoille River, is known by fly anglers for its annual *Hexagenia limbata* hatches. A public boat launch is located on the lake's south shore at Lake Caspian Public Beach in Greensboro.

Getting back to the Lamoille headwaters along VT 16, you'll soon see the Lamoille River rail trail that follows the road and a few small bridge crossings where you can access the stream. A stretch downstream from a bridge crossing on Riverside Farms Road, off VT 16 about 1 mile from the VT 16/VT 15 intersection, is promising. You won't miss the colorful Riverside Farms signage at the road.

After making a right turn at the VT 16/VT 15 intersection to follow the still small river downstream toward Hardwick, you'll pass a sizable road shoulder turnout directly next to the river with very nice fly-fishing water. You'll want to use 4- to 5-weight rods as you cast along the riffles and runs. A short distance beyond the road shoulder, you'll encounter a large parking area that serves the Lamoille Valley Rail Trail. A newly constructed, stately rail trail bridge crosses the river at this point. You can get a good view of the stream from the bridge. You'll notice a nice run upstream from the bridge crossing. The stretch downstream from the bridge opens into very wide water not nearly as nice as the upstream reach.

The rail trail, which will be your friend along the Lamoille for many miles, alternates sides of the river as it crosses back and forth at bridges along its course. Road shoulder parking areas continue to appear as you continue westerly on VT 15, following the river into Hardwick.

On its way to Hardwick and then into Wolcott, the river enjoys a stretch of faster moving water running directly next to VT 15 just before entering Hardwick. This is good rainbow trout water, appropriate for a longer rod than used farther upstream. VF&WD typically stocks nearly 3,000 rainbows between the falls in East Hardwick downstream to Pottersville Dam in Wolcott. Access at bridge crossings is popular with both conventional tackle anglers and fly fishers. Anglers target the area shortly after spring stockings, usually late May into early June.

After you get down into the Hardwick area along VT 15, other brooks that provide habitat for brook trout and a few small juvenile rainbows join Lamoille River. Cooper's Brook is an interesting tributary. The 2021 Vermont River Watershed *Basin 7 Tactical Basin Plan* showed that survey sampling in Cooper's Brook recorded 3,687 individuals per mile. These are undoubtedly mostly small annual wild brook trout. Cooper's Brook also took a beating from the July 2023 flood.

After exiting Hardwick you'll pass the dam as VT 15 crosses to the north side of the river. Just a few miles downstream from Hardwick, you'll reach the VF&WD Fisher Bridge Access parking area next to a historic covered bridge that once served as a railroad crossing (now part of the Lamoille Valley Rail Trail). The river becomes slow-moving water as it exits the covered bridge area. Admire the covered bridge then bypass this section after throwing a few casts along a run downstream from the bridge. Continue on VT 15 toward Wolcott for better fly-fishing water.

A nice fly-fishing stretch on a tributary that enters the river is located just before entering Wolcott. Elmore Branch has the reputation of being a spawning tributary for rainbow trout. The tributary itself is worth exploring at its mouth. Lamoille River on its southside near Elmore Branch's confluence is popular for fly fishing.

Before the devastating 2023 flood, Elmore Branch could be reached by taking Flat Iron Road off VT 15 to a recreational field access area. As of this writing, Flat Iron Road, wiped out in its section next to the river, is still blocked off. However, Elmore Branch and the stretches where it dumps into the Lamoille can be reached by taking a left onto School Street off VT 15. You'll reach the recreational fields in 0.4 mile after crossing the river. I performed a DNA sample 100 feet upstream from the mouth of Elmore Branch, and results showed 100 percent base-pair matches for very small amounts of both brook and rainbow trout. Fishing can be quite good both upstream and downstream from the Elmore Branch confluence. You'll notice a strong run that passes through old bridge abutments upstream from the mouth. The stretch downstream from the mouth flowing to the rail trail bridge is also popular.

WOLCOTT TO JOHNSON

The Wolcott Post Office is located just west of School Street on VT 15, downstream from the bridge crossing. Very nice rainbow trout water flows through small rapids and faster water both upstream and downstream of the post office. A footpath to the river next to an observation bench is located next to the post office. About 2 miles west of the post office, Corely Road intersects VT 15. Take a left onto Corley Road and drive to the small bridge crossing. A nice run is located a few hundred feet upstream from the bridge.

The Lamoille runs as a wide generally slow-moving river that picks up a couple of tributaries as it pushes downstream through Morrisville, Hyde Park, then Johnson, where it picks up the Gihon River. The Gihon is an important tributary in the area that's stocked with brown trout, typically 2,600 annually (see page 119). You'll notice the Lamoille's highly eroded stream banks and debris, a result of the 2023 deluge, as the river courses onward, picking up water from small tributaries. Tributary confluences, no matter how unproductive the entering stream itself might be during most times of the year, are always worth sampling during spring and again in fall. You'd be surprised what might turn up in your net. On your drive west toward Morrisville, you'll pass over the small Green River tributary, one of a few tributaries that attract spawning trout in this area.

A couple of access areas on the Lamoille in Morrisville are worth mentioning, although I don't find them particularly inviting. Directly across from the "Welcome to Morristown" sign along VT 15 heading into the village of Morrisville, you'll see formal VF&WD/Rotary access parking area signage. A footpath at the parking lot leads to stone steps that enter the river. The reach downstream from the parking area is better than the upstream, slow-moving water.

About 0.7 mile after the access lot, turn left onto VT 15A. You can get into the river at the bridge crossing. A good run flows just downstream from the bridge on its north side; another is located a short distance upstream from the bridge. Road shoulder parking is available a short distance up Darling Road next to the Lamoille Valley Rail Trail that runs along the river. Other than sampling those short stretches, it's best to bypass the area for better reaches downstream from Morrisville.

Some anglers like to try the water along Oxbow Riverfront Park in the village, located just 2 miles from the VT 15A bridge crossing. Continue on VT 15A to upper Main Street to a right on Portland Street (VT 100). Just around the corner, downstream from Oxbow, is a dam that creates Lake Lamoille. From VT 100, take a right onto Feline Loop to the dam area. I recommend bypassing this area after a look and instead heading a short distance downstream from Cady's Falls.

Take VT 100 to Cady's Falls Road then Duhamel Road. You'll notice Cady's Falls, an area of swift water at a second dam near the intersection. In less than a mile from the Cady's Falls Road/Duhamel Road intersection, you'll approach Cady's Falls Nursery on Duhamel Road, a gravel road. Be careful where you park, as the small lot along the road is intended only for trail users. However, easy parking is available along Duhamel next to the river. Just before reaching the nursery area, you'll see the small Kenfield Brook tributary. You can fish downstream to the brook's mouth, about 600 feet from its crossing on Duhamel Road. A short footpath at the bridge over the guardrail will get you into the small brook.

You'll encounter intermittent runs and riffles that alternate with long, slow-moving pools. Both hatchery and wild trout inhabit this reach. A variety of different mayflies and caddis pop up along this section throughout the season. I recommend an 8-foot, 4- to 5-weight rod and plenty of small streamers and dry flies here. I performed a DNA sample on Kenfield Brook during late April 2024. The results showed an enormous quantity of brook trout DNA and large amount of brown trout DNA; 100 percent base-pair matches, at lower values, were also found for rainbow trout DNA. My guess is Kenfield Brook is a healthy spawning tributary. I once ran into VF&WD fisheries biologist Lee Simanon over on the North Branch of Lamoille River and mentioned the substantial amount of salmonid DNA I had picked up during my sample. Lee mentioned that Vermont was about to acquire lands around Kenfield Brook as part of a FEMA buyout at the nursery. That section has been hit hard by high rain event flooding.

River stretches downstream from the confluence of Kenfield Brook run through a development called Ten Bends, operated by Lamoille Valley Property Owners Association, centered in the town of Hyde Park. The Ten Bends reach is popular with both fly anglers and recreational paddlers who pass through the area. The private mileage frontage allows fishing so long as only artificial lures are used (of no consequence to fly anglers). Ten Bends reach is operated as catch-and-release, but that rule doesn't really comport with official VF&WD regulations.

My understanding is that fishing access is still available via Ten Bends property a short distance from Duhamel Road, but I can't state that as absolute permission. The proper course of action is to check with the individual property owners if access is allowed via their private property. If access is allowed, the location is easy to find. From the Duhamel Road/Cady's Falls Road intersection, cross the bridge. In about 1 mile, turn left onto Main Street and in about 0.6 mile take a left onto Ten Bends Drive. You'll see a small "Fishermen's Parking" sign.

Take the road to other signage near the bridge that directs anglers to drive across a field to a small parking area near the stream. Check the mailbox at the designated parking area for any pertinent information or rules. During my last visit to the road, I chatted briefly with one of the residents, who confirmed that fishing is allowed by parking at the designated area at the far end of the field. But, again, check to make certain access is permitted. Always respect private property.

On your way to Johnson, downstream from Ten Bends, you'll want to sample the river along two roads: East River Road on the south side of the Lamoille just upstream from Johnson and Hogback Road, which hits a couple of reaches on the north side of the river. Sections of the Lamoille Valley Rail Trail parallel East River Road. The Lamoille also pours over Dog's Head Falls on its way to Johnson. Despite its scenic attributes, better fly-fishing stretches are available along East River Road and farther downstream on Hogback Road. (Hogback Road, by the way, will deliver you to North Branch of Lamoille River; see page 117).

Get onto East River Road by driving almost 4 miles west toward Johnson from the VT 15/Black Farm Road intersection in the New Hyde Park area, upstream. Take a left turn off VT 15 onto Railroad Street. Cross the truss bridge over the Lamoille, then take a quick left onto East River Road (Town Highway 46). After passing a few garage buildings and dwellings, you'll come to a large road shoulder parking area, and soon the Lamoille Valley Rail Trail joins the road. You can park next to the rail trail entrance. Formal rail trail parking is located at a point near Dog's Head Falls. The deep pool at the falls is popular with conventional tackle anglers. A run a few hundred feet downstream from the falls provides better fly-fishing water. You'll want to cast Woolly Buggers and small streamers.

If you continue past Dog's Head Falls, you'll eventually cross Waterman Brook (located 1.8 miles from the East River Road/Railroad Street intersection). The reach from Waterman Brook's confluence downstream to the rail trail bridge crossing is quite nice, as are runs and pools all along East River Road. This section is one of the few on the Lamoille River that has the appearance of a typical trout stream.

I performed a DNA sample on Waterman Brook a short distance upstream from its confluence with the river. The results showed interesting results. I had expected at least some rainbow trout (*Oncorhynchus mykiss*) DNA, since Valerie and I had observed a sizable rainbow darting up from the river in shallow water, right smack in front of our eyes that day in late March. A healthy presence of brook trout (*Salvelinus fontinalis*) was detected along with a very small quantity of brown trout (*Salmo trutta*) DNA.

PETERSON DAM REACH

Once the Lamoille River gets down as far as Peterson Dam in Milton, it has about 5 miles to go before it reaches Lake Champlain. The hydroelectric facility operated by Green Mountain Power has been under scrutiny by The Nature Conservancy. The Conservancy has identified the dam as having the highest ecological impact of Vermont's some 400 dams, a concern that's been held for years. A May 24, 2017, article in the *Milton Independent* reported that the dam blocks migration of landlocked salmon and endangered sturgeon from moving upriver. The Vermont Natural Resources Council advocated for removing the dam to help restore spawning habitat. In 2017 the dam, licensed until 2034, provided power to some 3,800 homes; stay tuned—that ongoing story that will undoubtedly resurface.

Peterson Dam in the third week of September. Foliage colors are just starting on the Lamoille River near Lake Champlain during that month. Landlocked salmon numbers increase once water level changes bring up more fish from the lake.

Hare's Ear and Pheasant Tail Nymphs *Tied by Bob Herson*

Freight Train *Tied by Dave McNeese*

This simple landlocked salmon and steelhead fly, based on a Randall Kaufmann pattern, is accented with purple hackle. Fluorescent orange/red wool, with black on the front, is ribbed with silver oval tinsel flash.

In the meantime, sometime between September and into November, fly anglers focus on the stretch downstream from the dam in search of landlocks moving in from Lake Champlain. Access is easy at the Peterson Dam Recreation Area off Peterson Road. Head to Milton, Vermont, then take US 7 to West Milton Road/Mayo Road then to Peterson Road. Parking is available at the dam at the end of Peterson Road. A wide footpath leads to the water just below the dam.

Fly-fishing guide Ralph Kucharek, who operates The Vermont Fly Fishing Company out of nearby Burlington, chimed in on fishing for Lamoille landlocks at the dam reach. Ralph also does a lot of landlock fishing on Lamoille and Winooski Rivers. "A word of caution about water discharges from the hydroelectric dam," he noted during our recent chat. "Water levels can rise very rapidly, so always be aware of water releases." A siren will alert those in the area. Don't waste any time getting out of the river when that happens. "I've been out there on water when it was coming up. You have to get out of there fast," Ralph said.

Lamoille River—North Branch

If fishing Lamoille River reaches around Hyde Park, you'll want to make a quick jaunt over to North Branch of the Lamoille River, located in the Waterville area. Stretches along these major tributaries to the Lamoille are quite picturesque, with covered bridges that provide a backdrop to runs and pools. The North Branch's character is much different than the main stem; it actually has the feel of a trout stream. Its geomorphology combines riffles, runs, and pools that provide nice fly-fishing water.

Born in a boggy area north of the hamlets of Belvidere Center, Belvidere Corners, and Belvidere Junction in Lamoille County, North Branch of Lamoille River flows west then south toward Waterville and its confluence with the Lamoille River, picking up small tributaries along its path, which roughly follows VT 109. Head to reaches upstream from Waterville for the best fly fishing. Waterville is located about 20 minutes northwest of Hyde Park and about 25 minutes northeast of Fairfax on VT 109.

You'll want to explore waters along and between several covered bridges that span the stream. The first covered bridge upstream from Waterville, Montgomery Covered Bridge, is off Montgomery Road. It's easily located off VT 109, about 5.6 miles northeast of the

North Branch of the Lamoille River at Montgomery Covered Bridge, located at the intersection of Montgomery Road and VT 109 in the Waterville area.

VT 108/VT 109 intersection. You'll find a convenient parking area located on the far side of the bridge. A footpath at the bridge leads directly to the stream.

The last time I was on the North Branch at the Montgomery Covered Bridge reach, I ran into VF&WD fisheries biologist Lee Simard laying out boundaries for a planned stream survey that was scheduled to follow in a few days. We took the opportunity during his lunch break to chat a bit about the North Branch. "All three species inhabit the North Branch," Lee said. "We stock brown trout, but the stream is also inhabited by rainbows and brook trout. Sizable browns inhabit some of the upstream pools."

After sampling the Montgomery Covered Bridge reach, move upstream to nice stretches along Back Road. On your way to Back Road, continuing north on VT 109, you'll pass Coddington Hollow Road and Jaynes Covered Bridge (0.5 mile from Montgomery Road). Back Road, about 2 miles beyond Coddington Hollow Road off VT 109, doesn't have a road sign. The road is on the left just after a bridge crossing, the next left turn after Smithville Road, and just past the white church building. Back Road follows the stream, passing over Lumber Mill Covered Bridge, then continues a little less than a mile to Morgan Covered Bridge off Morgan Bridge Road.

Park in road shoulder areas just past Morgan Covered Bridge, where you can get into the water before access becomes more difficult. In less than 2 miles, Back Road rejoins VT 109 at a bridge crossing. Not much direct access is available along this section of Back Road that passes by dwellings. Once back on VT 109, continue upstream to the Belvidere Center section, where it's possible to again access the water. You'll find a large, paved road shoulder turnout along VT 109 that provides access to the stream.

Gihon River

The Gihon River is born near Eden Notch at the Lamoille-Orleans county line, its tiny fingers nearly touching headwaters of East Branch of Missisquoi River, which flows northerly in an opposite direction. Its waters, along with other small streams, feed Lake Eden, an oddly shaped 186-acre recreational waterbody separated by two peninsulas. Lake Eden's waters dump over a spillway at its southwest end next to VT 100 and the hamlet of Eden Mills, creating what most consider the beginning of the Gihon River from a fly-fishing point of view. Eden Mills is about 10 miles northeast of Hyde Park. From Eden the river flows toward the village of Johnson, where it joins the Lamoille River.

Fishing access points are available at side road crossings off VT 100 and VT 100C, beginning in small water not far downstream from Lake Eden. Locate to the VT 15/VT 100C intersection in Johnson, then work your way upstream to small water that begins downstream from Lake Eden. While other areas are available to fish along the Gihon, the three described here are easy to locate along the lower, mid, and uppermost reaches.

The first access is located in Johnson on the lower reach, just 0.3 mile upstream from the VT 15/VT 100C intersection at the historic Power House Covered Bridge crossing on School Street. The original bridge was first constructed in 1870, decades before the power plant was built upstream. The bridge underwent a series of repairs until 2000, when sections collapsed from heavy snow loads. A new bridge was constructed in 2002. The old power plant structure was torn down in 2020, but the concrete foundational elements remain.

Other changes occurred at the site, of value to both fly anglers and other recreational users. The Beard family owned 600 feet of river frontage and waters downstream from the bridge, and swimmers and others were allowed to enjoy the site for many years. When

Headwaters of the Gihon River along the Blakeville Road reach. Blakeville Road marks the upper boundary of annual brown trout stockings by VF&WD. Fly-fishing sections are available along the Gihon River between Lake Eden and Johnson. VALERIE VALLA

Power House Covered Bridge in Johnson and the foundation remains of a hydroelectric power plant that once stood along the site. The structure was torn down in 2020.

the owners considered selling the 1.89 acres and river section, they worked with the Vermont River Conservancy and many local donors to create Beard Recreation Park to ensure continued access to the area. During summer you'll likely encounter swimmers and others using the picnic tables and small grills at the site, but at other times you'll have it to yourself while casting your streamers into the nice pools below the bridge. A small parking area is located on the far side of the bridge. Footpaths lead from the parking area down to the water.

Another nice fly-fishing reach, a bit more off the beaten track, is located upstream off Wilson Road. From the Power House Covered Bridge, drive 2.5 miles north on VT 100C to Wilson Road. You'll see Lamoille County Fairgrounds at the Wilson Road/VT 100C intersection. Take a left onto Wilson Road and drive less than a mile to a bridge crossing.

The Gihon River at Wilson Road bridge crossing. The Wilson Road reach is one of the nicest on the Gihon River.

You'll find plenty of pocketwater, runs, and small pools both downstream and upstream from the bridge. Park on the far side of the bridge along Wilson Road. Get into the water near the fire department water intake pipe at the bridge on Gihon Lane. A small road shoulder parking area is located 0.1 mile farther up Wilson Road beyond the bridge crossing. One option is to park down near the bridge and walk up to the area and then access the stream there. Fishing is also nice downstream from the bridge. It's fairly isolated aside from a couple of dwellings and structures along Gihon Lane and the stream. Sections upstream from the bridge eventually flow through isolated areas. Wilson Road eventually continues to the VT 100C/VT 100 intersection, near another bridge crossing. You'll want to move to this intersection by either continuing on Wilson Road or backtracking to the VT 100C/Wilson Road intersection then driving north 1.7 miles along VT 100C.

About 2.6 miles north of the VT100C/VT 100 intersection, you'll pass White Road and its small bridge crossing and a VT 100 crossing next to Eden Fields Lane just downstream from White Road. The bridge crossings are a short distance apart. This reach, 0.5 mile north of White Road on VT 100 (3.5 miles north of the VT 100C/VT 100 intersection), is worth a few casts on your way to the Blakeville Road stretch, located another mile up VT 100.

Blakeville Road is a small dirt backroad that runs just under 2 miles before it reconnects with VT 100. Not far after turning onto Blakeville Road, you'll pass over a small bridge crossing along the stream where you can access the water by parking on the road shoulder. A couple other access points are available along Blakeville Road before reaching another small bridge crossing with road shoulder parking, about 0.5 mile before the road reconnects with VT 100. The upstream side of the bridge accepts water that flows through well-shaded stretches.

I was interested in what species might inhabit the water upstream from the bridge. A DNA sample showed 100 percent base-pair match for brook trout. I was surprised that no brown trout DNA came through at that site. Caution should be made in interpreting whether the DNA was stream-born or came through the area from Lake Eden. Water upstream from the bridge is best fished with short rods and small nymphs. Downstream from the bridge, the Gihon flows through a more-open area, in some ways much easier to fly fish. You'll note masses of invasive, highly aggressive Japanese knotweed smothering the stream banks. Once you get beyond the bridge crossing on your way to VT 100, the stream flows next to dwellings where you might wish to stay clear of fishing.

Once you're back on VT 100, drive 0.4 mile north to the Whitney Lane bridge crossing. Park on the far side of the bridge. You can access the water next to VT 100 on the downstream side of the bridge. The stream at first exits beneath the bridge, dumping water into a nice yet brief small pool occasionally used by swimmers. The small brook-like water tumbles through an area that's heavily shaded. The water is somewhat difficult to negotiate using fly-fishing gear. Whitney Lane isn't far from the spillway at Lake Eden, although one more small crossing exists near Lake Eden Country Store.

Browns River

Browns River, a Lamoille River tributary, flows some 20 miles from its headwaters off the slopes of Mount Mansfield in the Underhill area to its confluence with the Lamoille in the town of Fairfax. The river at first flows westerly from its source then turns north, roughly following VT 128 until it reaches the Lamoille. Popular fly-fishing reaches can be found mostly around its uppermost sections near Jericho, Underhill, and Underhill Center at town parks and bridge crossings. If fishing the lower Winooski River corridor near Richmond,

Browns River at Old Mill Park, off Red Mill Drive in Jericho, is popular with naturalists, hikers, and anglers. The reach is accented with a trail system that follows the small waterfalls and pools.

a 15-minute drive north will get you to Jericho. If fishing the Lamoille River as far downstream as Cambridge, you can zip down VT 15 to the Jericho-Underhill areas.

While natural trout reproduction occurs in its upstream areas, Browns River also receives stockings of both brook and brown trout along three sections of the stream. Around 2,600 brook trout are stocked annually in its upper reaches between Stevenson Road in Underhill Center and Jericho Red Mill. From Jericho Red Mill downstream to VT 128, about 700 brown trout are stocked. The lowermost reaches, between VT 128 in Essex and Fairfax, receive 800 brown trout.

Easy access is available at Old Mill Park in Jericho and reaches upstream along Mills Riverside Park (located 2.9 miles west of Red Mill Road/Old Mill Park off VT 15). Mills Riverside Park and its 216 acres provide anglers, dog walkers, and others with stunning views of Mount Mansfield. Parking is available directly next to the footbridge that crosses the stream to the trail system.

If you're after brook trout, you'll also want to drive farther upstream into the Underhill Center area. Take River Road (located very close to the Mills Riverside Park entrance off VT 15) about 3 miles to Underwood Center, a small crossroads. You'll come to Moore Park on the left after crossing Mill Brook, a small tributary trickle that joins Browns River along the park grounds. A footpath at the small parking area leads to Mill Brook's confluence with Browns River. A pleasant streamside picnic table, placed in memory of Rose Mae Umberger, is a great place for a lunch break. A nice run flows along the area, perfect for your short 7-foot, 4-weight rod during late season.

After trying the Moore Park area, move upstream to Underhill Recreation Area, off Stevenson Road. Stevenson Road is a right turn off River Road, just upstream from Moore Park. You'll see the park located a short distance up Stevenson Road. The general recreation area, with its small swimming pond and tennis courts, has easy parking directly next to Browns River (actually more of a creek-size stream in Underhill Center). VF&WD fisheries biologist Lee Simard told me one of their stream sampling sites is located at a bridge crossing a short distance upstream next to Maple Leaf and Mountain Roads. While it's been some time since the river has been surveyed, their sampling revealed wild brook trout in Browns River's uppermost reaches. I took a DNA sample directly behind Underhill Recreation Center in November 2024. One hundred percent base-pair matches were found for both brook and brown trout. While I expected abundant brook trout DNA in this headwater reach, presence of a good amount of brown trout DNA confirmed a hunch. Perhaps brown trout spawning activity moved them that far upstream.

Missisquoi River

Termed a transboundary river because 15 miles of its 80-mile length flows through Quebec, Canada, before it turns and reenters Vermont, the Missisquoi River is known for both its coldwater and warmwater fly-fishing opportunities. Its name has been attributed to Abenaki origins: *Masipskoik*, meaning "where there is flint." The name is tied to flint chert formations near Missisquoi Bay.

The Missisquoi River is generally thought to first flow contiguously with Burgess Branch, originating from Belvidere Mountain southwest of Lowell. USGS topological maps show the Missisquoi formally flowing as a main stem river once Burgess Branch combines with East Branch of Missisquoi, a short distance northwest of Lowell.

Abundant brook trout can be netted in its headwaters and tributaries near its source northwest of Lowell in Orleans County and along the East Branch. Upper reaches of the main stem Missisquoi are heavily stocked for many miles along its length, as far down as the Enosburg Falls area Special Regulations Trophy Trout stretch, only 30 miles or so

The Loop Road reach, at a bridge crossing on the Missisquoi River main stem between Westfield and York, has nice runs and pools that harbor brown trout. VALERIE VALLA

from the river's terminus at Missisquoi Bay near Swanton. Trout fishing aside, one of the most remarkable fly-fishing opportunities is available in the river's lowermost reaches into Missisquoi Bay. A variety of warmwater species are prevalent and eager to smack your fly.

Missisquoi River—East Branch

The East Branch of Missisquoi River begins as a small brook-like stream a few miles south of Lowell near the Lamoille-Orleans county line. Once it passes Fiddler's Elbow Road, the little stream follows VT 100 for a short distance, picking up small amounts of water from small tributary inflows along the way. In less than a mile from Fiddler's Elbow Road, the stream shifts slightly to the west away from VT 100 and then roughly follows Mink Farm Road, crossing Cheney Road along the way. After another 4 miles or so, flowing north toward Lowell, it joins Burgess Branch, where it creates the main stem of the Missisquoi River. Before reaching its confluence with Burgess Branch, East Branch picks up the Ace Brook tributary and other inflows.

At the time of writing, VF&WD typically stocks 2,000 brook trout from Lower Village Road in Lowell upstream to along Mink Farm Road, but that could change in the near future based on recent electrofishing survey findings that show abundant brook trout inhabiting East Branch reaches upstream of Lowell. Survey data was compiled at a couple of locations a short distance downstream from Mink Farm Road and also upstream from Mink Farm Road near the Fiddler's Elbow Road/VT 100 intersection.

Around the time goldenrod starts showing in August, you'll want to get into those upper reaches; it's a great time to shed waders and wet-wade the soothing cold waters around Mink Farm Road and upstream. While the stream levels are low that time of year, you'll have little issue finding brook trout—albeit small in size.

Try the section off Cheney Road crossing, located off Mink Farm Road about 2.5 miles south of Lowell. Then relocate upstream to the VT 100 crossing next to Fiddler's Elbow Road. Park on Fiddler's Elbow Road at the bridge crossing where it joins VT 100 but be sure to pull

Cheney Road crossing, off Mink Farm Road, in August. Although water levels are low, brook trout can be caught along small pools and runs in the cool water. VALERIE VALLA

East Branch of the Missisquoi departs the VT 100 bridge crossing at Fiddler's Elbow intersection, picks up water from an unnamed tributary, then flows through a wide-open area flanked by goldenrod in August. The heavily shaded section upstream from the bridge crossing has abundant young-of-the-year brookies. Good spawning habitat exists through the spruce stand and woody deadfalls that overlie gravel beds.

way over to the side. Large trucks come through there heading to the quarry up the road. A large culvert where you'll park transports water from an unnamed small tributary. The plunge pool below the culvert is typically full of small brook trout. Use small flies, preferably barbless. Most of the brook trout are small young-of-the-year that are anxious to eat your fly, yet a few dollar bill–size fish can be in the mix. The inflow meets the main channel of the East Branch in just a hundred feet or so, where it dumps in just downstream from the VT 100 crossing. You'll devote most of your time on the main channel.

At the VT 100 crossing, the East Branch flows through a concrete box culvert. The small plunge pool on the downstream side of the bridge always has eager yet small brookies. After departing the culvert and picking up additional water from the unnamed tributary, the East Branch flows through a wide-open area flanked by goldenrod and brush. Brook trout inhabit the small pools.

Stretches upstream from the bridge crossing are much different. The small water flows through a heavily shaded stand of spruce. You'll be doing a lot of crouch fishing, almost dabbling in the small pools, with little casting involved. VF&WD biologist Lee Simard chimed in about brook trout numbers along that reach, the focus of headwater population survey sample stretches. "This is another of our sample sites—some of the highest brook trout numbers I've ever seen in a run, albeit mostly young-of-the-year. Lots of downed trees, great gravels," Lee said.

Incidentally, if you were fishing East Branch of Missisquoi River in late season, a 10-mile drive up and over Hazen Notch Road (VT 58) in Lowell, near the Burgess Branch confluence, would put you on the upper Trout River and its tributaries at Montgomery Center. The Trout River is a Lamoille tributary. Keep in mind that Hazen Notch Road is almost entirely a seasonal dirt road, closed during winter into early spring. Even when it opens to travel after winter, mud season and poor road conditions can make this route a poor choice. Instead you'll have to take an alternate, much longer route over paved roads to reach Montgomery Center.

This small brook trout was caught on a size 18 Lime Stonefly at the Fiddler's Elbow on the East Branch in mid-August. Many of the brookies are young-of-the-year size, but a mixed age-class inhabits these upper reaches.

Missisquoi River—Main Stem

LOWELL/WESTFIELD AREA TO TROY NEAR THE CANADIAN BORDER

Mid- to late May is a great time to sample the Missisquoi River's main stem headwaters, located not far north of Lowell, downstream from the confluence of the East Branch with Burgess Brook in the Westfield area. Biologist Lee Simard provided information from longitudinal water temperature surveys on the East Branch and miles downstream on main stem tributaries. His Stream Temperature Monitoring report, compiled from 2022–2023 studies, revealed observations that any fly angler visiting the lower and upper Missisquoi reaches would find interesting. His longitudinal study was conducted downriver on The Branch, a tributary that enters the Missisquoi downstream from Enosburg, and along East Branch headwaters and the Missisquoi downstream from the Burgess Branch–East Branch confluence to the Lowell-Westfield town line. The study emphasized the importance of maintaining good tributary habitat protection and enhancement for the vitality of the Missisquoi River.

"While the East Branch Missisquoi River is suitable for high-quality trout populations, the Missisquoi River is likely too warm, especially for brook trout, largely due to warming caused by the Burgess Branch. Although the Burgess Branch watershed is largely forested, the closed Belvidere Asbestos Mine, a hazardous waste site, likely produces runoff with elevated water temperatures that transfer downstream. Tributaries farther downstream of this survey, including Snider Brook, Taft Brook, Mill Brook, and Lily Brook among others, all have high-quality trout populations, especially at higher elevations. Habitat work in this portion of the watershed should focus on restoration and protection of these tributaries rather than along the main stem river itself."

Lee's study also pointed to Ace Brook, which enters the East Branch south of Lowell, as a tributary that "appears to contribute substantial warming" of the system, as does Burgess Branch. There's much more to the report. Get a copy of VF&WD document F36R25Study10-41 to read the entire report.

Your first stop should be at the Loop Road off VT 100 between Westfield and Troy. By the time the Missisquoi River clears Loop Road, it has picked up Snider, Mineral Spring, and Le Clair Brooks not far upstream from Loop Road. At the VT 100/Loop Road intersection, drive 2 miles to the bridge crossing. Park along the shoulder on the far side of the bridge. The upstream section has nice runs that flow toward the bridge, while the stretch downstream from the bridge is flat water. Brown trout swim in both areas. Explore areas upstream from the bridge to a point where Mineral Spring Brook enters the river. During late May, keep your eyes open for fish rises along the entire reach.

After a stop at Loop Road crossing, you'll want to get to River Road in Troy. Sections along River Road are varied in structure, often flowing well away from the road, and are some of the most interesting for sure before the Missisquoi vanishes into Canada in North Troy. About 0.9 mile from the VT 100/River Road intersection, you'll hit the first bridge crossing. My suggestion is to skip this reach that encounters a small dam; it's not that fabulous for fly fishing, although bait casters enjoy fishing at the bridge pool since parking isn't an issue, and trout undoubtedly inhabit the water after spring stockings. Instead, put effort into other areas along River Road downstream.

A nice section at the Veilleux Road crossing, about 3 miles past the last bridge crossing, has a nice run that's stocked with brown trout in May. The Veilleux Road crossing was once blessed by an iconic 111-year-old covered bridge that was tragically destroyed by

A nice stretch downstream from the Veilleux Road crossing. Swing small streamers through the run, especially in late May. Brown trout inhabit the reach. VALERIE VALLA

fire on February 6, 2021, when a snowmobile caught fire while crossing the bridge. Swing streamers through the run, located a couple hundred feet downstream from the crossing.

At about a mile past the Veilleux Road crossing, you'll encounter fast water along River Road, just before the Missisquoi drops through Big Falls. Extreme caution should be exercised if you're wading this stretch. If you're getting into the water at all, sample the first manageable run at the beginning of the fast water and stay clear of other sections—in my opinion it's too dangerous to be wading that water. You can park along the road shoulder. You'll find significant Yellow Sally (*Isoperla*) stonefly flights during mid-May. Try small bead-head flies in this run, then head farther downstream below the falls

One of the most scenic and impressive areas on the upper Missisquoi is along the large pool at Big Falls. If you're thinking about getting a fly deep enough into the pool itself where the falls crash in, make sure it's not during summer. Other recreational users, especially swimmers, enjoy the easily accessed pool. Big Falls State Park lands, with 2,400 feet of frontage, was donated to the state by Citizens Utilities Company in 1996.

Just past the fast water on River Road, you'll see an entrance into the park. Continue driving north past the entrance about 0.3 mile. You'll see the dirt Town Access road on the left. In early season, the gate might be closed. If it is, walk around the gate and along the river. You might encounter fish rising along the far bank of the slow-moving river as you make your way up the dirt road toward a small utility facility building. Reach the falls by walking a couple hundred feet to the left of the building. Signage welcomes visitors to the area. Locals report taking large brown trout in the deep pool on conventional tackle or worms.

River Road ends just over a mile north of the Town Access road, where it enters North Troy. Just downstream from the North Troy bridge crossing, visible from the road, the

Big Falls along River Road dumps into a large pool popular with both anglers and other recreational users, especially swimmers. The deep pool can be difficult to negotiate with fly-fishing tackle. More manageable stretches flow just around the bend.

Missisquoi River reaches a dam at the Missisquoi River hydro facility then pushes downstream less than 2 miles, where it crosses into Canada. After spending a short time in Canada, it reenters the United States at East Richford.

EAST RICHFORD US-CANADA BORDER CROSSING TO SWANTON DAM

The Missisquoi River resumes its journey to Lake Champlain in East Richford at the United States–Canada border. VF&WD typically stocks 2,400–2,800 brown trout from the border downstream to the VT 118 crossing in East Berkshire, where the Missisquoi picks up the Trout River.

Beginning at East Richford along VT 105A, you'll quickly see nice runs almost immediately downstream from the border, about where the tiny Lucas Brook tributary enters the river. This reach has some of the nicest water along this section. Always pay attention to areas where tributaries enter the river, no matter how small—especially in early fall. In a couple of miles, VT 105A merges into VT 105, the road that follows the river in and out of sight all the way to Enosburg Falls.

In about 2.5 miles from the VT 105A/VT 105 intersection, VT 105 passes under railroad tracks. Just before entering the narrow overpass, you'll see a small parking area that provides access for both anglers and recreational floaters. Canoers enjoy the relatively new access point, constructed by professionals and volunteers from the Northern Forest Canoe Trail (NFCT) group in the summer of 2023. Funding was provided by the Upper Missisquoi and Trout Rivers Wild and Scenic Committee. The Missisquoi was designated a Wild and Scenic River in 2014.

Stone and timber steps lead directly into the river. The access point is the first available after exiting the international border and allows anglers to float through a picturesque area on the river for many miles past islands and Class I–II stretches. During spring and early season, as well as periods of higher water, the river is tough to wade at the East Richford NFCT access point, given its deep pools and runs. Later, when river levels drop in fall, you can wade-fish the runs downstream from the railroad bridge crossing.

Floating is probably the better way to fish the long reach between East Richford and Richford Village. Angling floaters can take out in Richford at the Richford Conservation and Access Area easement. Class III ledge rapids not appropriate for float anglers continue past the easement area and then run into the Davis Park NFCT access, located off River Street.

As you continue driving along VT 105 into Richford, you'll come across a couple of road shoulder parking areas near the river, but these are scant. VT 105 skirts Richford, where it gathers the North Branch of Missisquoi that comes in from Canada. The pools here are wide, large, and difficult to wade during early season. Canoers and kayakers use the river access at Davis Park. Signage welcomes users to the park's 2 acres along the Missisquoi River. When water levels allow, you can locate at Davis Park then approach the tail out of the rapids, located just upstream.

The Missisquoi pushes into East Berkshire downstream from Richford, where it gathers the Trout River just upstream from the VT 118 bridge crossing. You'll want to take a quick diversion up the Trout River and explore its numerous tributary brooks. Brook trout are stocked in areas that provide quick access. Before entering the Enosburg Falls area, about 1.5 miles from the VT 118/VT 105 intersection in East Berkshire, VT 105 crosses the Missisquoi Valley Rail Trail. A small rail trail parking turnout is next to the trail crossing along VT 105. The rail trail closely follows the river here along an interesting section spattered with exposed bedrock runs and riffles that negotiate small islands. The water here warms significantly during summer, making it more appropriate for smallmouth bass fishing. The Enosburg Falls Special Regulations Trophy Trout stretch is just a couple of miles away.

Popular with both fly fishers and conventional anglers when Enosburg Falls receives ample rainbow trout stockings along the Special Regulations mileage, the reach also warms significantly in summer. If you're interested in sampling the area, do so during mid- to late May. VF&WD typically stocks between 1,000 and 3,000 (depending on the year) large rainbow trout between 15 and 16 inches.

While 5.6 miles of river occupy this section that runs downstream from the falls to Kane Road, concentrate your efforts around Enosburg Falls, casting streamers through areas of runs along exposed bedrock. Start at the falls, located off Duffy Hill Road near the striking "Bridge of Flowers." Parking is easy along the road turnout. A footpath from the parking area leads down a short hill and then along the river as it turns away from the road. You'll be in the company of conventional tackle anglers, who enjoy landing nice-size trout. The stretch that runs away from the road for a short distance downstream from the falls also has good structure. Once the river departs Enosburg Falls, it again flows as mostly flat water without much structure all the way to Kane Road. The river again gains structure about 0.5 mile downstream from the Kane Road crossing, but the reach is difficult to access unless you have a way to float it. Parking is available at the Kane Road bridge.

After you're down the river that far, I'd recommend heading to Swanton, located about 17 miles from Enosburg Falls. You might encounter interesting fly fishing at Swanton Dam, less than 10 miles upstream from Missisquoi Bay. Note the white posted sign at the river, reminding anglers that the water from the top of the dam downstream approximately 850 feet is off limits to fishing from March 16 through May 31 to protect sturgeon spawning waters. Buoys mark the restricted area. Red signage posted downstream from the buoys

Enosburg Falls area supports a Special Regulations Trophy Trout stretch that runs about 5.6 miles from Enosburg Falls Dam downstream to the Kane Road bridge crossing. Much of the mileage is neither prime fly-fishing water nor easily wadable.

Swanton Dam, in Swanton Village, attracts anglers who target a multitude of fish species that wander upstream from Missisquoi Bay on Lake Champlain.

alerts anglers that the section is restricted from March 16 to the first Saturday in May. Check the official regulations for any updates or additional restrictions.

Locals and anglers who frequently fish the dam, easily accessed on both sides, will tell you that almost any species that lives in Lake Champlain might be encountered, depending on the time of year. Access the dam via Marble Mill Falls Park at the bridge crossing in Swanton or the formal VF&WD boat launch site on Foundry Street on the opposite side of the river. At the opposite side of the bridge, take Foundry Street to the launch site. Park at the launch parking lot and walk along the river upstream to areas that are conducive to wade-fishing.

VF&WD stocks muskie fingerlings at the dam and also downstream in Missisquoi Bay in cooperation with New York State's muskellunge hatchery, located at Chautauqua Lake in western New York. It's part of a well-publicized effort by both states to restore this native species to Lake Champlain. You'll also want to experience warmwater fly fishing on Lake Champlain's Missisquoi Bay area itself.

Lake Champlain North Section at Missisquoi Bay Area

I had never witnessed a carp's willingness to chew a fly until one day in late May when I received a message from Drew Price: "Get up here on Tuesday, the fish are in!" The destination was an area around Missisquoi Bay, well north of the areas off Otter Creek where I got my bowfin.

That message could have meant most any Champlain species "are in," since Drew targets everything swimming among the emerging cattails near the shoreline. Within minutes we saw bowfin, although my neglecting to thoroughly set the hook deep led to the fish's escape. Carp were in the area too. Drew spotted a big one. "Not that big, but I'll get him," he said.

Carp and other species hang among emerging cattails and weed growth in the backwaters around Missisquoi Bay.

Captain Drew Price, Lake Champlain's premier fly angler and guide, runs his boat to the weed beds around the Missisquoi Bay area in search of many different warmwater species. After arriving at a destination, Drew carefully poles the boat through the shallows where carp, bowfin, bluegill, bullheads, and other fish hang out.

Lake Champlain's Drew Price battled this big carp, which he spotted among others on a late day in May in the Missisquoi Bay area. The fight took him through gobs of weeds before the fish came to the net.

Drew has a different definition of "not that big." That fish, approaching 35 pounds or so, was quite a catch. Even more amazing was Drew's confidence that he could get the monster out from under the vegetation. I would have just clipped my leader and moved on. After 10 minutes or so, the fish was netted. "What leader strength do you have on, Drew?" I asked. He smiled and laughed, "12-pound test—non-fluorocarbon." Later we targeted bullhead, gar, and largemouth bass. Despite having to deal with a monstrous thunderstorm and heavy rains, after the storm passed Drew got the bass he had already targeted. The double rainbow we saw was a testament to an outstanding day on Champlain's waters with fly rods.

Shaggin' Draggin' *Tied by Pat Cohen*

Pat Cohen's go-to carp fly

Whisker Biscuit *Tied by Pat Cohen*

Pat's Whisker Biscuit carp fly will entice carp anywhere. Pat ties the pattern on a #8 short-shank nymph hook. The body is crafted from Pat's own Carp Dub collar, laser dub body. Large bead chain eyes get the fly down to the fish.

Jay Zimmerman's Carp Bug (variation) *Tied by Bob Adams*

Trout River

A dendritic complex of small streams and brooks converge to form the Trout River, an important gateway stream for trout running up from the Missisquoi River seeking spawning areas. A cursory study of USGS topography maps of the area reveals a number of small waters that converge around Montgomery and Montgomery Center, a few miles southeast of East Berkshire, where the Trout River joins the Missisquoi. Among the waters that feed Trout River from the north are Black Falls, Hannah Clark, and Jay Brooks. Wade Brook and the South Branch of Trout River flow northerly to Montgomery Center.

While Trout River receives a significant number of hatchery brook trout stockings from the VT 118 bridge in East Berkshire to River Street in Montgomery annually in late May—nearly 2,000—some of the more interesting fly fishing can be had along upper reaches near Montgomery Center. There's lots to explore, although some reaches require effort to access.

If you're after stocked trout, there's no easier access than River Walk Park, known more as a brook trout put-and-take reach. However, you'll want to fish that section in May or early June. You can expect company after the put-and-take stockings take place. Summer months bring plenty of swimmers. The park is located 4.2 miles south of the VT 118/ VT 105 intersection in East Berkshire, at the VT 118 bridge crossing. You'll encounter a couple of quaint covered bridges on the way to River Walk Park.

Better opportunities for wild trout exist well upstream in the Montgomery Center area tributaries and the South Branch of Trout River. Montgomery Center is a great little town, known for its small inns, pubs, and other streetscape pleasantries. Valerie and I have enjoyed staying in the town when we're in the area. You'll want to get your breakfast at a neat little place called Bernie's, located right on Main Street, before heading out on water.

An important tributary of the Missisquoi River, Trout River has a good population of wild brook trout in its upper reaches and along its numerous tributaries. The easiest access to Trout River is directly next to VT 118 at River Walk Park in Montgomery, a popular area to fish for stocked brook trout from late May into early June.

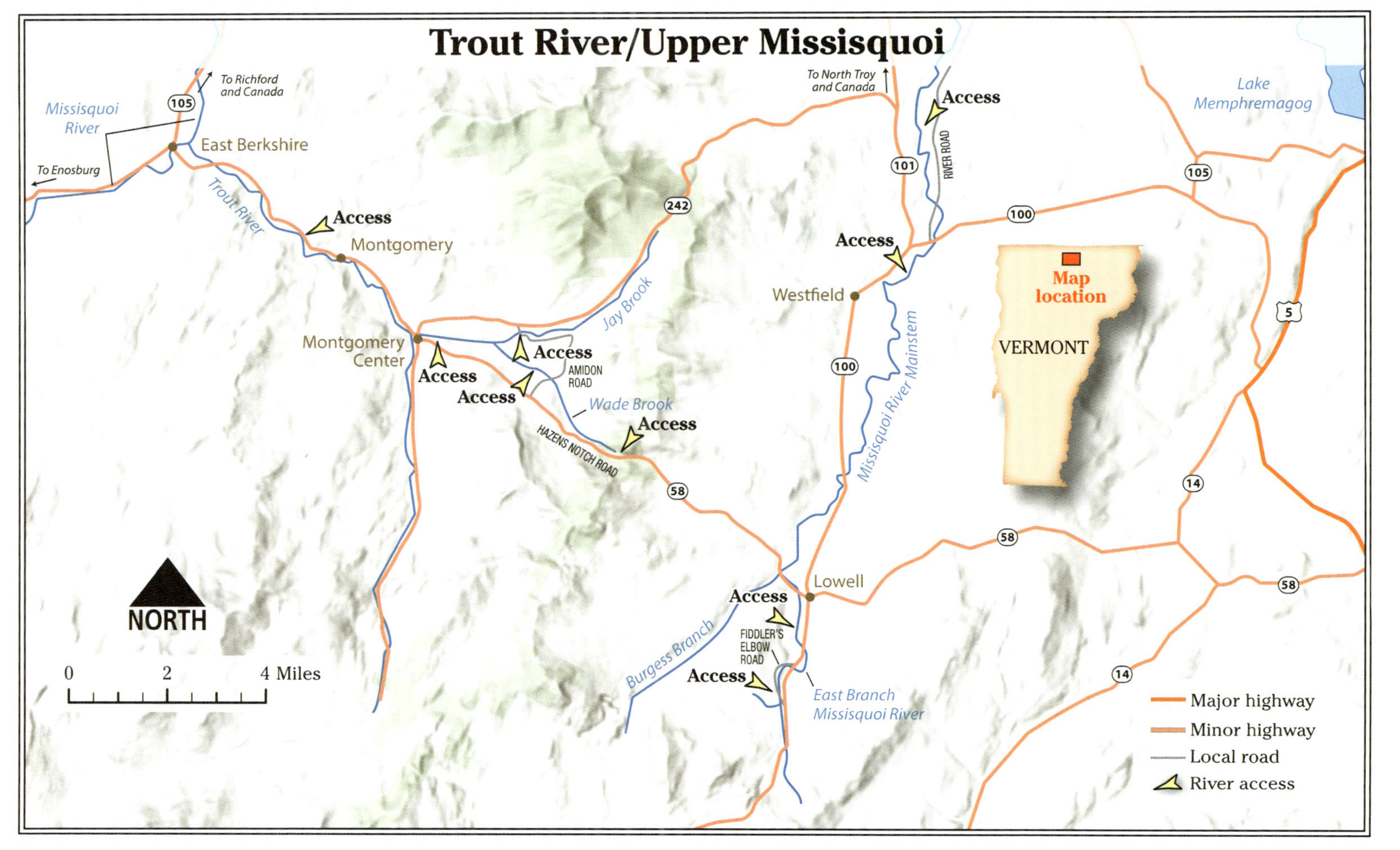

Trout River/Upper Missisquoi
To Richford and Canada
105
Missisquoi River
East Berkshire
To Enosburg
Trout River
Access
Montgomery
Montgomery Center
Access
Access
Access
AMIDON ROAD
Wade Brook
Jay Brook
242
Access
HAZENS NOTCH ROAD
58
To North Troy and Canada
Access
101
RIVER ROAD
Access
Westfield
100
100
Missisquoi River Mainstem
Map location
VERMONT
Lake Memphremagog
105
5
14
58
58
14
Lowell
Access
FIDDLER'S ELBOW ROAD
Access
Burgess Branch
East Branch Missisquoi River
NORTH
0
2
4 Miles
Major highway
Minor highway
Local road
River access

Small and isolated yet pleasant tributary reaches that flow into Montgomery Center provide splendid fly-fishing experiences for those who value smaller yet very beautiful wild brook trout and some brown trout.

I ran into VF&WD biologist Lee Simard over on the North Branch of Lamoille River in summer 2024 while he was laying out site work for an upcoming trout survey. We chatted a bit about lots of things, including the Trout River and its tributaries, including the South Branch of the Trout River, located not all that far away over in the Missisquoi watershed. I stayed in touch with Lee after our encounter and further discussed Trout River. "While the Trout River itself has pretty poor wild trout populations, most of its tributaries are pretty good. Others of note from our sampling are Black Falls Brook and Jay Brook [not to be confused with Jay Branch, which flows down the east-facing side of the mountain by the ski resort down into the Missisquoi main stem in Jay/Troy]. Numbers are okay higher up but not great. Also, Hannah Clark Brook [is pretty good]."

Trout River Tributaries—Wade Brook and Jay Brook

While the other small Trout River tributaries provide really nice brook trout water, I've confined my fishing in an area I call "the triangle," a section around Montgomery Center bounded by Amidon Road to the east, VT 242 to the north, and VT 58 on the south. You'll be able to sample two nice tributaries within the triangle: Wade Brook (my favorite) and Jay Brook.

VF&WD has a couple of survey areas on Wade Brook, both showing abundant brook trout. From Main Street in Montgomery Center, take VT 58 (Hazen Notch Road) close to 3 miles to Amidon Road. Amidon Road is a dirt road beset with mud issues in early season, but by the time things dry out it's very drivable. From the VT 58/Amidon Road intersection, drive 0.5 mile to the Wade Brook bridge crossing.

Wade Brook at the Amidon Road Crossing. Wade Brook provides excellent fishing in its small pools and runs that support wild brook trout and a few wild brown trout. VALERIE VALLA

Jay Brook, easily accessed off Amidon Road, is known for its small brook trout and a few brown trout, but not in the same numbers that typically inhabit Wade Brook.

Park along the road shoulder. A footpath is located next to guardrail that at first is pretty good but quickly turns into a very rocky "trail" just before reaching the stream. You'll note the very large-diameter galvanized culvert that transports the brook beneath the road. Solar panels are mounted on the top surface, no doubt to facilitate water level monitoring. The small water pockets and pools are best fished with short rods and small flies that entice brook trout and a few small brown trout. If you had stayed on Hazen Notch Road and driven another few miles beyond Amidon Road, you would have encountered a small concrete bridge crossing on a Wade Brook tributary. This area also serves as a VF&WD survey sample section. You can park just down the road from the bridge crossing, at small lot reserved for snow plows.

Your next stop off Amidon Road will be access into Jay Brook. Continue past the Wade Brook crossing about 2 miles; Jay Brook will come into view as your approach a bridge crossing at the corner of Amidon and Montgomery Heights Roads. The area has dwellings here and there, including one downstream from the bridge. You can get into the water at the bridge area. In less than 0.5 mile, Amidon Road intersects VT 242. Montgomery Center is only 2.2 miles from the intersection. (If you want to sample Jay Brook first, take the VT 242 route from Montgomery Center, a much shorter distance than via Hazen Notch Road to Amidon Road.)

Once you're back in Montgomery Center, you'll be on the Trout River. Wade and Jay Brooks combine upstream, creating Trout River and interesting reaches right in the center of town, at the VT 118/VT 242 intersection bridge crossing. Just upstream from the bridge crossing, a series of pools and runs create a gathering place for both locals and visitors, and often anglers. Swimmers and partyers hang out along the stream's lower pool, accessed just a short distance up Hazen Notch Road. (You would have passed it on your way to Amidon Road.)

A short trek down the steep hill off the road will get you into the area called a few different things, even by locals who grew up in the area. One local who has lived his entire life in town said the entire series of pools is collectively called "Third Hole." Others call it "The Holes" or "First, Second, Third Holes." Parking along the road shoulder at two separated areas is extremely limited. You'll not want to even try to fish it during summer afternoons, when it's always busy with other users. But the water looks fabulous.

Incidentally, when seasonal Hazen Notch Road is open to vehicles, a 10-mile drive from Montgomery Center will get you to the Lowell area and the headwaters of the Missisquoi River and its East Branch. Otherwise, a much longer route over paved roads will deliver you to those watersheds.

Trout River—South Branch

South Branch of Trout River has abundant brook and a few brown trout swimming in its waters. Lee Simard has completed survey samples along South Trout River as far up as about 5 miles south of Montgomery Center, at a VT 118 pull-off rest area (concrete picnic table area). "We do get a few brown trout in the South Branch, although it is predominantly brook trout. The past couple years sampling at the concrete picnic table area has just been brook trout, but we have gotten a few 8-inch browns in the past," Lee said.

South Branch of Trout River headwaters, located about 5 miles south of Montgomery Center, is small water during late summer yet remains cold enough to support lots of small brook trout. You'll want to use short rods and approach the small pools and runs with stealth. VALERIE VALLA

Small yet nice looking brook trout inhabit South Branch's headwaters. Recent population surveys completed by VF&WD along upper reaches show abundant brook trout inhabiting the water.

Water levels along that reach are very low during summer but still quite cold. Brook trout inhabit the small pools and runs. You'll be doing lots of crouch fishing that time of year, using short rods and small flies. In summer 2024 I took a DNA sample out of curiosity that a few brown trout might be in the mix. The results showed, not surprisingly, extremely high values for brook trout DNA but none for brown trout.

Moving downstream from the concrete picnic table reach, your next stop should be at Gibou Road, 3.3 miles north of the picnic table site, off VT 118 (1.6 miles south of VT 242/VT 58 in Montgomery Center). Drive to the bridge crossing, once the site of a historic 55-foot lattice truss covered bridge that spanned the South Branch going back to 1899. The bridge was torn down in 2002, but you'll see the old abutments upstream of the modern bridge that now spans the stream. Take a few minutes to observe and read about the covered bridge at an official State of Vermont historic marker. It's located directly next to a shoulder pull-off where you can park.

A well-worn footpath leading from the road shoulder at the new bridge will quickly get you into the beautiful pool created by water tumbling through rock formations. You'll no doubt encounter swimmers at the pool during summer's heat, but at other times of the season you'll have it to yourself. Should you encounter other users at the pool, just move upstream by walking around the plunge area. As is true upstream, this reach is predominantly brook trout water, with the chance of encountering brown trout here and there.

Gibou Road reach at the gorge area, the old site of the Hectorville Covered Bridge crossing. The pool is a popular gathering place for swimmers seeking refuge from summer's heat. During cooler summer days or during spring and fall, fly fishing can be enjoyed in solitude. VALERIE VALLA

The city of Newport is home to Lake Memphremagog (in the background) and Clyde River landlocked salmon runs. VF&WD stocks thousands of smolt landlocked salmon annually in the Clyde River near its mouth at the lake. After growing in the lake, they return to the Clyde River, to the joy of many fly fishers.

Memphremagog Basin

Island Pond, a small community on the shores of Island Lake in the Northeast Kingdom, is a launch area that can put fly fishers on a number of interesting trout streams, ponds, and numerous lakes. The basin has upward of nearly 60 lakes and ponds 10 acres or larger. Vermont's 2023 Department of Environmental Conservation *Basin 17 Tactical Basin Plan* reported that most lakes and ponds in the basin support migratory fish that run tributaries to spawn. Among the most notable are Lake Memphremagog; Willoughby, Crystal, Salem, Seymour, and Echo Lakes; and Norton, Island, and Great and Little Averill Ponds. A 10-minute drive west will deliver anglers to Seymour and Echo Lakes, two waterbodies that afford some of the best lake trout and landlocked salmon fishing in all of Vermont. Jobs Pond, a Special Regulations brook trout water, is just 5 minutes away; Lake Willoughby is just 14 miles away. Within quick distance from Lake Willoughby (and its namesake Willoughby River), Bald Hill and Long Ponds are just around the corner too. The Pherrins River, a Clyde River tributary, flows almost within view of Island Pond itself. The city of Newport on the shore of Lake Memphremagog also serves as a launchpad for the lower Clyde River and other waters north of Island Pond, such as Echo Lake and Lake Seymour.

Memphremagog Basin includes a wide variety of waters that support a wide variety of gamefish. Tandem smelt patterns, such as this Willoughby Smelt Tandem tied by noted New Hampshire streamer fly tier Scott Biron, are effective on deep, cold lakes such as Lake Willoughby, Echo Lake, and Lake Seymour. Smelt are an important forage fish for lake trout and landlocked salmon.

Front-and-center rivers that flow through the basin include Willoughby, Black, and Barton with their annual steelhead runs out of Lake Memphremagog. The Clyde River, noted for its lower-reach landlocked salmon runs and its upper reach for sizable brook trout, is among the most interesting. Pherrins River is also a favorite among brook trout anglers.

Pherrins River

Pherrins River's nearly 10-mile journey to its confluence with the Clyde River at Island Pond begins near the area of Norton's Pond, tucked well away into Vermont's extreme northeast. Its southerly flow passes through a variety of habitat ranging from boggy areas to stretches shrouded with overhanging trees and cover, perfect for its wild brook trout. While access points are easy at a few points along VT 114, wading among its bowling ball–size rocks and often mucky pools can be difficult. Pherrins River has the reputation of supporting sizable brook trout, although dollar bill–size and smaller finger-length brookies are quite abundant.

Some of the more isolated reaches not frequently fished, and not well publicized, are probably best left for those willing to explore on their own. The river's reputation as a healthy brook trout fishery is by no means a secret, since sections of very good water are located along VT 114 bridge crossings. I'll describe just two common yet nice access points that will get anglers new to the Pherrins started.

The Pherrins River at a VT 114 bridge crossing, about 2.5 miles north of the VT 105/VT 114 intersection in Island Pond. VALERIE VALLA

Island Lake, in the community of Island Pond, is a multispecies fishery inhabited by trout, bass, and other gamefish.

From the VT 105/VT 114 intersection in Island Pond, drive 1.2 miles north on VT 114 to a bridge crossing. Five Mile Square Road joins VT 114 at the bridge on the left, Iron Bridge Loop on the right. Park along the road shoulder on Five Mile Square Road. (Five Mile Square Road leads to a formal VF&WD Clyde River access popular with canoe floaters.)

From the parking area, get into the river at the bridge, then fish upstream to and beyond the railroad trestle that can be seen from the bridge. You can also fish downstream from the bridge, but note that the stream flows past dwellings both upstream and downstream. Respect private property as you make your way along this reach. During my younger years, I wouldn't have needed a wading staff to negotiate the stream, but using one will make your experience easier, whatever your physical ability.

After casting along this reach—I use a 7-foot, 4-weight rod—continue upstream to another access point 1.2 miles north of the Iron Loop Bridge crossing, on VT 114. Park along the large road shoulder at the bridge crossing (it's large enough to accommodate a few vehicles, even an RV). The parking area is right next to the railroad tracks that lead to downstream areas.

You can fish the stretch upstream from the bridge by quick access into the stream. Keep in mind that in early season, this can be more difficult water to negotiate and fish than in late season. The water stays plenty cold throughout the summer, to the delight of both brookies and anglers, I'm sure. You'll find yourself ducking overhanging branches while working upstream. As you continue driving upstream along VT 114, about 0.3 mile from the large road shoulder parking area, VT 111 intersects VT 114 on the left. Turn left onto

Iron Bridge Loop area off VT 114, close to Island Pond, has excellent brook trout habitat both upstream and downstream from the bridge.

This brook trout was enticed at Pine Brook during low-water conditions. The brook is quite tiny and requires negotiating cramped space.

VT 111 and you'll come to a bridge crossing almost immediately. Try that stretch before heading upstream to an interesting tributary named Pine Brook.

Pine Brook is very tiny water. However, it's full of little brook trout. Get back onto VT 114, then continue north 2 miles to a one-lane concrete railroad underpass next to a marshy, boggy area where you would expect to see a moose. Use caution as you pass under the railroad crossing. Large trucks and other vehicles frequently zoom toward the one-lane underpass with minimal regard to safety. From the underpass, drive another 0.8 mile to guardrails that mark the point where tiny Pine Brook flows through two large culverts beneath VT 114. To park for access to Pine Brook, it's best to head up VT 114 a short distance then turn and head back to Pine Brook, where you can barely park a vehicle on the far side of the bridge. Use caution exiting your vehicle while stopping to have a look at the diminutive brook.

The dual culverts that dump water could stand some alterations to allow brook trout to venture farther upstream. The small brookies eager to strike small dry flies congregate at the spillways. Most are small, but there are dollar bill–size fish in the mix. After exiting the culverts, the brook flows a short distance to the railroad overpass that can be viewed from the road. If you have back problems, you're likely to exacerbate the condition from crouching and bending as you make your way through the underpass. The small plunge pool can be fished by cramming yourself under the short height space beneath the railroad. You'll need a very short rod, too, as you negotiate the brook, working downstream through overhanging tree limbs.

Norton Pond is a warmwater fishery in the neighborhood of the Pherrins River. Fly anglers target pike that inhabit its water.

NORTON POND AND COATICOOK RIVER

While you're in the area of the upper Pherrins River, two other waters worth fishing are in the neighborhood, off VT 114. Norton Pond's VF&WD boat launch is located 9 miles north of Island Pond and about 7 miles west of Canaan, just south of the Canadian border. Norton Pond, a warmwater fishery, is also popular with paddlers. Fly anglers target pike and bass. Its outlet, Coaticook River, is inhabited by wild brown trout and wild brook trout. Access the Coaticook on Gagnon Road, off VT 114. From the VT 114/Gagnon Road intersection, drive about 0.2 mile to the small bridge crossing. You can get into the stream there. A few nice runs are located downstream from the bridge.

The Coaticook River, Norton Pond's outlet, is inhabited by brook and brown trout. Access the stream on Gagnon Road, off VT 114.

Jobs Pond

Jobs Pond has been well studied by VF&WD biologists. This special resource has received attention to devise the best management practices to ensure a healthy natural brook trout population, balanced with expectations from anglers who fish the pond. VF&WD fisheries biologist Jud Kratzer kindly provided information from his July 1, 2022–June 30, 2023, Salmonid Inventory and Management report.

> Vermont has ten ponds where wild Brook Trout abundance is high enough to provide angling opportunities. Most of these ponds are remote, but some are easier to access. Jobs and Martins Ponds are the only wild Brook Trout ponds where a trailered boat can be easily launched. These ponds are also known for producing large Brook Trout. A creel survey on Jobs Pond suggested that a large proportion of the Brook Trout population was being harvested by anglers (Kratzer, 2013). With the goals of increasing Brook Trout abundance and identifying Jobs and Martins Ponds as unique fishing opportunities, a test waters designation was implemented in 2013 at Jobs and Martins Ponds. During the test waters designations (2013 to 2019), anglers were only permitted to harvest two trout per day, and there was a protected slot limit. Anglers could not harvest Brook Trout between 305 and 381 mm (12 and 15 in.), and not more than one harvested fish could exceed 381 mm (15 in.). One of the main objectives of this test waters designation was to increase angler catch rate of Brook Trout larger than 305 mm (12 in.). The test waters designation was replaced with a permanent regulation in 2020. The current regulation sets the daily creel limit at two, with no length restrictions.

One of Vermont's nicest natural brook trout waters is Jobs Pond, a real gem located up the road from Island Pond. Jobs provides good fishing for brook trout during fall season, as well as in spring. VALERIE VALLA

Besides its abundant brook trout population, the pond is favored by anglers because of its easy access and watercraft launch area. From the VT 105/VT 114 intersection in Island Pond, drive 2.5 miles to Newark Road. Signage on Newark Road will direct you to the pond. An informational kiosk with pond fishing regulations is located at the parking area next to the pond. You can unload your floating craft next to the pond and then park your vehicle at the designated parking area.

Mickey Finn Bucktail *Tied by Mike Valla*

Brook trout love hitting this striking pattern, most effective during early-season pond fishing. They are best fished by slow trolling relatively close to the shoreline.

Fishing tactics vary on Jobs Pond, just as they do on most other small trout waters. During early season, and again in early fall, brook trout will cruise closer to shore than during summer. Note that fishing is allowed from the second Saturday in April until October 31. During mid- to late April, an effective strategy is to slowly troll streamers along the shoreline. Mickey Finn bucktails are a good choice. During late June into very early July, you'll likely run into a *Hexagenia* hatch, something I wasn't previously aware of.

Leighton Wass, who recently wrote that wonderful book *Fly Fishing the Hex Hatch*, tipped me off during a time I was looking for a *Hexagenia* pond I had not yet fished. "Did you fish Jobs?" Leighton asked. "There *is* a Hex hatch there, and a few huge wild brookies in there." Fall is also a magnificent time of year to fish Jobs. From the vantage point of the launch area, you'll view an absolutely amazing display of fall foliage colors and breathtaking beauty. *Hexagenia* fishing aside, if asked what time of year I would suggest fishing Jobs, I would instantly recommend sampling Jobs in the fall, largely because of its sheer splendor at that time of year.

Bald Hill Pond

Bald Hill Pond, one of a few ponds within a quick drive of Lake Willoughby and not far from Jobs and Island Ponds, lies in a setting that provides a sense of inner peace if not rainbow trout in the net. The public boat launch provides easy launch of your kayak or canoe. Streamers slowly trolled up and down the shoreline out of a canoe can excite trout earlier in the season. Bald Hill Pond lacks the depth enjoyed by other ponds in the area, such as Long Pond. Maximum depth, in the pond's center, is only 42 feet. Early-season fishing is best.

Dirt roads lead to the pond no matter what direction you approach from, and they can be quite nasty just after mud season. Once things dry out a bit, a small sedan can easily negotiate the dirt roads that lead to the launch. If approaching from Lake Willoughby, take Newark Pond Road off VT 5A, the road that runs along the east shore of Lake Willoughby. Signage marking Newark Pond Road is not apparent; keep your eyes out for the dirt road that enters VT 5A.

If approaching from Jobs Pond in the north, you'll take Abbot Hill Road to Newark Pond Road to Bald Hill Road. You might want to check out Vermont's Bald Hill Fish Culture Station, located at 66 Abbott Hill Road in Newark. You'll pass it while making your way to Newark Pond Road. Created in 1952 as a salmon and brown trout hatchery, today the facility raises a variety of gamefish: brown, rainbow, and steelhead trout, walleye, and landlocked

Bald Hill Pond, a rainbow trout fishery, is located a few miles southwest of Jobs Pond and due east of Lake Willoughby. A formal VF&WD boat launch is at the pond. VALERIE VALLA

Governor Aiken Streamer *Tied by Scott Biron*

This Vermont classic is a perfect style streamer to troll for rainbows during spring.

salmon. It's the only hatchery raising Atlantic salmon broodstock. Eggs are supplied to other federal and state hatcheries.

Newark Pond Road eventually runs smack along Newark Pond, a warmwater fishery that also has easy access. Pick up Bald Hill Pond Road where it joins Newark Pond Road, just beyond signage that marks the Newark Pond Natural Area, located at the north end of the pond. The formal boat launch is about halfway up the east shore of Bald Hill Pond. An information kiosk is located at the launch. Bald Hill Pond is stocked with the new Eagle Lake strain of rainbow trout. Anglers are asked to report catches to the VF&WD.

Long Pond

Long Pond isn't that far from Bald Hill Pond as the crow flies. However, driving to it can be confusing if you're approaching from Bald Hill Pond. You'll want to head toward the Westmore area, on the northeast end of Lake Willoughby, as a starting point. Long Pond is just 2.5 miles from Westmore. Whether approaching from Westmore or from points at the south end of Lake Willoughby, take Long Pond Road off VT 5A. The intersection is located at Willoughby Lake Store on VT 5A. Take Long Pond Road to the VF&WD parking area and information kiosk.

Like Bald Hill Pond, a VF&WD boat launch provides quick access into the pond. A kayak or canoe will serve you well. Unlike Bald Hill Pond, Long Pond receives annual stockings of brook and lake trout but not rainbow trout. Long Pond is also much deeper

Long Pond, northwest of Bald Hill Pond, is another interesting waterbody in the vicinity of Lake Willoughby. The deepest part of the pond is about halfway between the boat launch and the small island that's in view from the parking area.

than Bald Hill's modest depths. The deepest part of Long Pond, about 75 feet maximum, is about halfway between the launch and the small island. Water on the opposite side of the island gradually drops off from 50-foot depths to the pond's shallowest area at the extreme south end. I've been told that a spring area exists on the far-left side of the pond and for that reason can be productive. I have not fished that end myself. If you're on the pond during late June or into July for evening fishing, your fly box should have a few *Hexagenia* patterns, as that bug emerges during that time of the season.

Lake Willoughby

I see it's a fair, pretty sheet of water,
Our Willoughby! How did you hear of it?
I expect, though, everyone's heard of it.
—Robert Frost, "A Servant to Servants"

Even before Robert Frost became famous for his poetic gift, he was inspired by Lake Willoughby, a waterbody that captured his imagination. Frost and his family vacationed on the lake that has been called one of the most beautiful in all of New England. While the salubrious environment around the lake captured Frost's imagination, its fishery captures the imagination of anglers who have "heard of it," knowing that large landlocked salmon and rainbow trout swim its waters.

More than 300 feet deep in some areas, Lake Willoughby is one of the deepest lakes in New England. Its depth is second only to Lake Champlain, which figures in at 400 feet. Fly anglers fish the lake itself, employing different strategies. There are times when tandem streamers work well, trolled with downriggers. Ted Benoit, who operates Lead & Tackle

Lake Willoughby, located east of Barton and southwest of Island Pond in the Northeast Kingdom town of Westmore, has been called one of the most beautiful in New England. Its fishing also holds charm.

Vermont Rainbow Smelt *Tied by Scott Biron*

Tandem streamers work well on Lake Willoughby most any time of year. Smelt inhabit the large, deep oligotrophic lake.

Co. outfitters in Lyndonville, carries tandem streamers trolled on Lake Willoughby tied by local tier Al Pitt, who operates Wild Branch Fly Shop in Lunenburg. Ted had this to say about when to troll streamers on Willoughby, and how to do it:

"It's always a good time of year to troll a streamer," he explained. "It's very effective for browns close to the surface in shallow water in early spring. Also, midsummer, when the water heats up and the thermocline sets in, you can effectively troll them at 30–50 feet. Trolling flies can be super-effective bait on the surface or at depth. I find that they look more natural than a spoon and cut through the water more calmly than an erratic lightweight spoon."

Ted went on to explain why trolling streamers is so effective. It has to do with the forage base. "These clear clean lakes and ponds in the Northeast Kingdom are loaded with baitfish like dace, shiners, and smelt. 'Matching the hatch' becomes figuring out which color baitfish the fish will be homed in on that time of year. Being able to accurately hold your bait at a depth is key to locating fish and finding success while trolling. These tandem flies will track nicely and attract fish at any depth. Speed can be adjusted to target different species with these same flies. While lakers could be more aggressive with a slightly slower presentation, rainbows are so excitable while feeding that they may like a faster-pace troll and be more apt to chase a faster moving bait. Same can be said about salmon. Many people will troll spoons or flies at 2–3 mph, making adjustments for the wind, chop, and species targeted. I always find fishing is better with a good chop on the water."

Willoughby Willie *Tied by Scott Biron*

Other streamers, like this Willoughby Willie, can elicit strikes from trout and landlocked salmon when they're cruising close to the water's surface.

Trolling streamers deep on Lake Willoughby and other lakes in the Northeast Kingdom requires a way to get down to the fish. Ted considers Keith Chamberlain of Burke, Vermont, one of the most knowledgeable streamer-trolling anglers in the area who spends a significant amount of time on Lake Willoughby. "Keith invented a downrigger device that can be adjusted to different pound release tensions and is what will release your line from the downrigger cable when a fish is on."

Regarding best practices, several factors will change bait and presentation speed/depth, according to Ted. Among the variables are the time of year and how aggressive or passive fish will be: "What types of bait are in the body of water? How much bait is there? Are fish homed in on bait or bugs? Rainbows, for example, are bug eaters. That is why you can find them near the surface in 100+ feet of water during midsummer.

"However, they also eat smelt and other baitfish and will be found at depths of 40+ feet, as well feeding on other resources. It just goes to show how healthy some of these fisheries we have are. These fish have many different food sources all year round. That's why fishing in places such as Lake Willoughby can be so difficult. It's so clear, and there is so much natural food."

Yet, as Ted pointed out, there are also times when trout and landlocks are closer to the surface and equally attracted by streamers trolled with sink-tip lines. Both contemporary and classic streamers are effective; some patterns tied to suggest the rainbow smelt that inhabit Willoughby's waters are commonly used on Lake Willoughby and other lakes and ponds in the area. Dry-fly fishing can also take trout at times when they swim near the surface, cruising for food that includes mayflies.

Typically during the second week of July, according to Leighton Wass, more than a few fly anglers anticipate the annual *Hexagenia* hatch on Willoughby. Leighton, whom I got to know pretty well, and on one occasion fished with during the Hex hatch on Seyon (Noyes) Pond, aptly described *Hexagenia* emergences on the various ponds and lakes he fishes as having a "rolling opening." The smaller, shallower ponds are first, with *Hexagenia* arriving around late June. A "last hurrah," as Leighton said, happens on Willoughby. I asked him about Lake Willoughby's Hex emergence timing and variability:

"The Hex on Willow [sic] is the last one to appear in Vermont because the water stays much colder than other lakes because it's so deep. It used to be that I could count on it starting around mid-July. It has been a week earlier these past years, but the way things are going right now, the 15th sounds about right for this year, depending on the weather between then and now.

"There are enough salmon in Willow so that there is always a chance of hooking into one of them as well. And perch. Ugh! And rock bass. Double ugh! All the 'bows are wild. The salmon, for the most part, are stocked fish."

Once the hatch is on, Leighton pointed out, all kinds of gamefish feast on the bug; even suckers are suckers for the juicy bugs, both adults and emergers, on Willoughby. You'll hook into most anything at times, not just rainbow trout or landlocks. Just as was described on Seyon Pond, when fishing Willoughby, Leighton rigs a rod with a sink-tip line. In his book *Fly Fishing the Hex Hatch* (2022), Leighton described a scenario on Willoughby when smelt were breaking the surface, obviously being chased by a gamefish. A single-hook smelt streamer did the trick. Fishing "pre-hatch," just as on other Hex waters, you'll also want to fish nymphs of various styles. As described in the Seyon Pond section, Kennebago Muddlers are a good choice. But a wide variety of emerger patterns work well too.

Willoughby River

The lake's outlet, Willoughby River, also causes a stir for fly anglers, especially during the spring steelhead run. Willoughby River exits Lake Willoughby's north end, flows toward Brownington Center, and then runs along Brownington Center Road. Before it reaches Orleans, the river makes a dip in direction then continues, roughly following VT 58 (Willoughby Avenue). Just before Willoughby Avenue intersects East Street, you'll encounter a large parking turnout on the right that provides access into the stream with a little effort involved. While you'll encounter juvenile rainbows along this stretch during summer, keep in mind that Special Regulations are in effect earlier in the season. Fishing is closed from the Orleans/Brownington Road bridge to the top of the natural falls from March 15 through May 31.

After crashing over Willoughby Falls, the stream eventually joins the Barton River. The Barton River flows north to Lake Memphremagog. More than 10,000 juvenile steelhead are stocked annually during mid-April in the Willoughby River between Brownington Center Road bridge and the Barton River confluence. The supplemental stockings join juveniles from natural spawn and head downstream to Lake Memphremagog, where they spend 1–4 years growing before heading back up through the Barton River to Willoughby River.

Willoughby River along Brownington Center Road, a few miles upstream from Orleans and Willoughby Falls.

Willoughby Falls in Orleans attracts not only steelhead but also spectators hoping to catch a view of the fish leaping the falls. A formal VF&WD parking lot is located at the bridge crossing off East Street in Orleans.

Rainbow trout eager to migrate upstream during spawning runs jump up Willoughby Falls. Chances of viewing leaping rainbows are best in late April. COURTESY VF&WD

Various egg patterns, including Estaz Egg flies, should be in your fly box when fishing any steelhead river, including the Willoughby.

Some head into the Black and Clyde Rivers to spawn. Most fish are around 18 inches and 2 pounds, but fish up to 4 pounds are caught every year.

The steelhead run, particularly at Willoughby Falls, is one of the most anticipated natural events in the area. Local diner and store clerks in Orleans will tell you about it should you mention you're an angler. Steelhead can be viewed leaping, or trying to leap, up the short falls just outside of Orleans. Non-angling spectators frequently gather at the falls to view the show. A formal VF&WD parking lot at the bridge crossing on East Street provides easy access to the falls. An information kiosk is located at the lot. As is true when fishing any stream that receives annual steelhead runs, you'll have plenty of company while casting your favorite steelhead flies.

If you want to view a leaping rainbow, head to the falls during late April. VF&WD fisheries biologist Jud Kratzer had this to say: "I expect some rainbow trout to start running the Willoughby River in late March, but they usually don't start jumping until late April. The flows and water temperatures have to be just right for them to jump out of the water, and that usually happens the third or fourth week of April."

Shadow Lake

If visiting the Willoughby Lake area in early July, know that *Hexagenia limbata* is likely to be emerging on Shadow Lake. The lake, located near Glover, is a half-hour drive or so southwest of Lake Willoughby. Head to Glover on VT 16 (it's south of Barton). Stop by Curriers Market in Glover for sandwiches and drinks before continuing south on VT 16 just short of 5 miles to Shadow Lake Road. Turn right onto Shadow Lake Road. You'll drive past a beach area before reaching a formal VF&WD boat launch parking lot. A launch "greeter" is on staff at the lot to wash your small craft and examine it for any evidence of milfoil or other invasives. The lake has been free of milfoil for several years.

The 210-acre, 139-foot-deep lake is stocked annually with lake and rainbow trout. Local anglers who've own cottages on the lake for many years told me the lake doesn't fish as well for rainbows as it did in years past. However, one angler I spoke with wasn't sure if it could be attributed to differences in fish stocking levels over the years. One fly angler who owns a cottage on the lake suggested that if rainbows are the target species, head to the inlet at the west end of the lake.

While you might hear reports of *Hexagenia* arriving on smaller ponds to the south in late June, Shadow Lake straggles a bit. Valerie and I made two separate trips to Shadow Lake during late June, dragging our two Hornbeck canoes with us in anticipation of getting into a hatch. We were about two weeks or more early but had a wrong hunch—that because of the unseasonable hot spell, Shadow Lake and Caspian Lake to its south would experience an earlier than usual hatch. To avoid another unproductive trip, I finally resorted to leaving my phone number with the launch greeter at the parking lot, who kindly offered to contact me when the bugs arrived.

Shadow Lake, located south of the Lake Willoughby area in Glover, is known for its annual *Hexagenia* hatches.
VALERIE VALLA

Echo Lake

If fishing near the Lake Willoughby area or surrounding ponds, check out Echo Lake. Access into its waters is easy. VF&WD operates a public launch on the south end, at 264 West Echo Lake Road. An information kiosk is located at the launch. The launch is usually staffed by a local friendly official who will check your private craft for invasive plants. Fed by neighboring Lake Seymour's outlet, Echo Lake is a deep oligotrophic lake with a maximum depth of 129 feet. Echo Lake's outlet flows into the Clyde River, then Lake Memphremagog. Stocked and wild salmonids swim its water, joined by a few warmwater species that include smallmouth bass.

Routinely monitored by state fisheries biologists, VF&WD stocks rainbow, steelhead, and lake trout. Biologist Jud Kratzer provided useful information about Echo Lake's fishery. The 2022–23 *Inland Fisheries* report he prepared explained rainbow and steelhead stockings in Echo Lake. The purpose of an ongoing study is to determine the best management of rainbow trout stockings in deep, cold lakes in Vermont.

"In 2022, the Department began evaluating a new Rainbow Trout strain for stocking in Vermont," Jud wrote. "The Eagle Lake strain of Rainbow Trout was chosen because of its good performance in the hatchery and in the wild in other states. It is being compared with Erwin-Arlee for culturability in the hatchery and catchability in put-and-take stocking situations. In waters that are being stocked with Rainbow Trout on a maintenance basis, the Eagle Lake strain is being compared with the Willoughby Steelhead strain." While conclusions are still under study, it appears that the Eagle Lake strain might be the better choice for rainbow trout maintenance in Echo Lake and others.

While early-season streamer trolling is effective on Echo Lake for both lake and rainbow trout, you'll not want to ignore the *Hexagenia* hatch during early to mid-July. I asked longtime Vermont *Hexagenia* fishing authority Leighton Wass his take on Echo's fishery

Echo Lake, northwest of Island Pond, is one of two deep and cold lakes in the Clyde River system; the other is Lake Seymour, its connected sister waterbody. VALERIE VALLA

and Hex emergence timing. He has netted some dandy fish during that hatch on Echo Lake and many other Vermont ponds and lakes.

"Echo comes on about the same time as Caspian. There are *huge* lake trout in there, and there used to be *huge* rainbows. But in recent years, Echo has fallen severely," Leighton said. His records typically show Caspian Lake and Echo's Hex hatches come on around July 1. Like other Hex hatch waters, you'll often see a few popping the surface in late afternoon, but as Leighton says, it doesn't mean much to the angler; patience is a virtue. Often the blitz doesn't even start until just before 9 p.m. Just as on Seyon Pond and others, rig two rods—one with a sinking tip for subsurface fishing prior to the hatch with nymphs or emergers.

Of course, you're better off finding a place to camp or stay within relatively close distance of Echo Lake, especially if you're planning a late evening *Hexagenia* outing. White Birch Lodge, located on the east shore between Echo Lake and Lake Seymour, is ideal. However, only four cabins are available at this very nice, family-owned lodge. Other options are campgrounds near Lake Willoughby.

Governor Aiken Tandem Streamer (Jim Warner variation) *Tied by Scott Biron*

Jim Warner (1928–2015), iconic streamer fly tier from Hampshire who was well-known by fly fishers who trolled streamers on Vermont's ponds and lakes, created this variation of the Governor Aiken pattern. Jim added beads to his version of the tandem variation because, according to Scott Biron, he felt the beads would push water to help suggest the swimming action of smelt and other baitfish.

Plenty of other streamers also work well on Lake Echo, depending on the time of year you're on the water. Try the same patterns that work well on nearby deep lakes and ponds in the Northeast Kingdom, such as neighboring Lake Willoughby. Ted Benoit, who operates Lead & Tackle Co. outfitters in Lyndonville, carries interesting streamers in his shop, given their effectiveness on several waters. Among them is Al's Little Perch and Copper Kettle, tied by Vermonter and commercial fly tier Al Pitt of Lunenburg. Ted had this to say about those two streamers:

"Both of those flies would be awesome for about anything the Northeast Kingdom has to offer. The copper is a go-to color with the clear/clean water conditions we have here in the

Al's Little Perch *Tied by Al Pitt*

Al Pitt's Little Perch streamer is another good choice if trolling on Echo Lake. Along with smelt, yellow perch inhabit the lake.

Copper Kettle *Tied by Al Pitt*

The same hook configuration as in the Al's Little Perch is used in this flashy streamer. The body is created with Copper Holo Braid.

Lake Seymour's outlet feeds Echo Lake. Lake Seymour is a deep and cold oligotrophic lake that supports both coldwater and warmwater species. Lake trout, landlocked salmon, and bass are targeted by anglers. A formal VF&WD boat launch is located off VT 111, just 6 miles from Echo Lake's launch site. *Hexagenia* mayflies also emerge on Lake Seymour.

Northeast Kingdom. Lots of guys will use copper/silver/gold to imitate local baitfish and as a 'go getter' attraction spoon. Depending on the time of year, things are always eating perch in both the small ponds and big lakes like Willoughby. We have great luck up there with red/orange flies in general, especially in the fall. This is a great time of year to be on the water. The best in my opinion!"

I asked Al Pitt his take on the Copper Kettle streamer, knowing myself that copper flies can be very effective on New York state waters, especially fished on ponds and lakes. "The Copper Kettle seems to me to be a fly that is usually only fair at best—*but* when the fish want copper, it can be great."

Clyde River

CLYDE RIVER—LOWER RIVER LANDLOCKED SALMON SECTION

When fly fishers mention the Clyde River, it's almost always in the primary context of the river's 1.5-mile or so reach between just downstream of the Number 1,2,3 dam at Clyde Pond and Lake Memphremagog in Newport—the landlocked salmon reach. The awesome landlocked salmon that run the Clyde from Lake Memphremagog far up in the Northeast Kingdom in Newport is only half the story. The Clyde is a tale of a fishery that once flourished, then declined, then found new life. It's a story filled with biological and corporate utility company misdeeds and controversy, yet tempered with positive corrections taken place in more recent times. At one time it seemed as if the river commanded attention from every special interest group, ranging from anglers to fisheries biologists and culturalists to power companies to conservation activists to environmental attorneys to journalists—to even the Vermont State Supreme Court. All surrounding that magnificent gamefish revered by fly anglers everywhere—landlocked *Salmo salar.*

The very entity that contributed to the river's decline, reaching back many decades ago, today devotes vast amounts of effort monitoring and enhancing the Clyde's salmon runs. VF&WD is definitely committed to the Clyde and has done an outstanding job addressing the monumental task of bringing the runs back to sustained levels to be enjoyed and benefited by all. Yet during the 1930s and into the 1940s, VF&WD trapped salmon running the Clyde to obtain eggs for hatcheries, something that undoubtedly affected the fishery.

Scott Wheeler's *When Salmon Was King* (2007) is a must-read book for any fly fisher planning a trip to the Clyde River. Wheeler describes Clyde River salmon fishery impacts that occurred years ago. Annual egg-stripping activity is commonly mentioned as a factor that led to the landlocked salmon decline. In Wheeler's pages we learn that in 1933, a high of 500,994 eggs were stripped from landlocks at a station in the Clyde's waters; a low of 101,251 eggs were harvested in 1943. While overharvesting eggs during those years seems a plausible reason for negatively affecting the landlocked salmon, it's also possible that other factors could have contributed to the fishery's decline. However, most agree that a dam located a short distance upstream from Newport was a major contributor to the Clyde's salmon run declines. Number 11 Dam on the Clyde, constructed in 1957, stopped salmon in their annual upstream journey from Lake Memphremagog.

The Number 11 dam is gone now. Both Mother Nature and human intervention had something to do with demolishing what was once a major obstruction to fish migration. A high rain event and spring runoff in 1994 occurred first, which led to a dam breach. At first the utility company operating at the dam quickly tried to rebuild it. But the State of Vermont's legal interventions, as well as involvement by the Northeast Kingdom Chapter

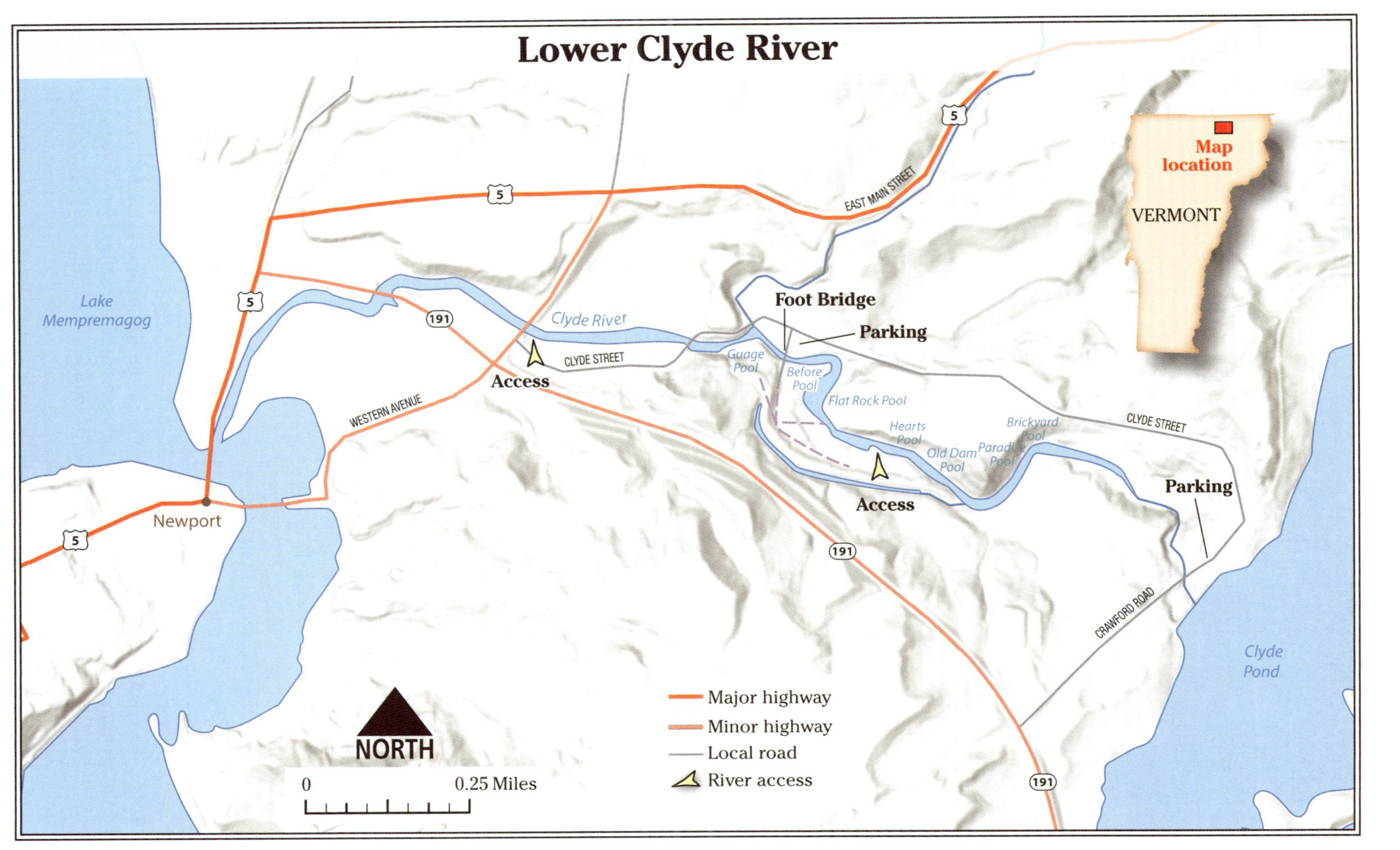
Lower Clyde River
Map location
VERMONT
5
EAST MAIN STREET
Lake Memphremagog
Foot Bridge
Parking
191
Clyde River
CLYDE STREET
Access
Guage Pool
Before Pool
Flat Rock Pool
Hearts Pool
Old Dam Pool
Paradise Pool
Brickyard Pool
CLYDE STREET
WESTERN AVENUE
Access
Parking
Newport
CRAWFORD ROAD
Clyde Pond
Major highway
Minor highway
Local road
River access
NORTH
0
0.25 Miles

Jeff Newhouse landed this Clyde River landlocked salmon in mid-October. Every fall, landlocks run up from Lake Memphremagog in Newport. Anglers from throughout the Northeast converge on the Clyde River during that time. PARKER WRIGHT

of Trout Unlimited and others, quashed the plan. Completely dismantling the dam resulted in additional upstream stretches for the salmon to move through.

Meanwhile, as part the Federal Energy Regulatory Commission (FERC) relicensing of the Number 1,2,3 powerhouse at Newport and dam at Clyde Pond still operating required construction of fish-passage facilities. A fish ladder constructed in 2007 at the Great Bay Hydro Dam powerhouse off Clyde Street in Newport allows for holding salmon then trucking and transferring fish to upstream Clyde River locations. That action along with other fishery-enhancing practices have been tried, including comparing different strains of landlocked salmon—the Sebagao strain and the Memphremagog strain. Stockings are required to maintain the fishery and angling opportunities. Thousands of smolt-size juvenile salmon are annually released into the Clyde at Newport.

Today the Clyde at Newport attracts anglers from throughout the Northeast. Parker Wright, a longtime guide at the Fly Rod Shop in Stowe, is an enthusiastic Clyde River fly angler. Parker had this to say about the Clyde's salmon runs: "The Clyde is a well-known river for good reason. Every September it gets a run of landlocked Atlantic salmon coming out of Lake Memphremagog. These salmon are some of the largest in all of New England

and will put even the most seasoned anglers to the test. It's hard to beat swinging a streamer 6 inches below the surface on a crisp autumn morning and see a 3- to 6-pound landlocked salmon come out of nowhere and hammer the fly. When the streamer fishing is slow, bust out a nymphing rod and some small BWO nymphs and get to work!"

Anglers interested in fishing the Clyde's lower river landlocked salmon reaches should head to Newport, a lovely city on the shore of Lake Memphremagog. You'll see a few anglers on the water as early as late August. Expect lots of company during salmon run season, especially in mid-October. (Anglers wanting an early-morning breakfast before heading out to entice landlocks should try The Brown Cow, located at 350 East Main Street. The small friendly diner is open at 5:30 a.m.)

While plenty of enthusiastic fly anglers will be roaming the waters, access to the river is quite easy. In Newport, make your way to the VT 5/Clyde Street intersection. You'll immediately find parking and quick access to the river. Again, expect others to occupy the space. Continue 0.2 mile on Clyde Street to a bridge crossing, then keep your eyes open for a very small formal VF&WD parking area on the right in another 0.2 mile. A short footpath leads across a small bridge. Once you cross the bridge, you'll walk right smack into an information kiosk that includes a rough map of all the pool names and their locations. The first pool you'll encounter as you make your way up upstream along the footpath from the small bridge is Before Pool.

During a Clyde outing the third week of September, I ran into Paul Bugeja on Before Pool. Paul is an enthusiastic landlocked salmon angler known for his role as chairman of the Vermont Chapter of the Native Fish Coalition. As a member of the NFC, I can't say enough good things about their work to "Protect, preserve, and restore wild native fish populations through stewardship of the fish and their habitats."

Vermont Native Fish Coalition Chairman Paul Bugeja casts a muddler minnow on Before Pool during a third-week-of-September outing on the Clyde River while Rosalina checks his progress. Paul uses a variety of patterns while fishing for landlocked salmon on the Clyde.

The Clyde River stretch between Flat Rock and Hearts Pools. The water level was unusually low when this photo was taken in late September 2024.

Paul calls annual Clyde River landlocked salmon trips the fall pilgrimage. "You'll see fly anglers here from all over the Northeast this time of year," he said. "They come from Rhode Island, Connecticut, and other areas to get a chance to experience catching a landlocked salmon."

Paul had already tallied three landlocks during the 2024 season by the time I saw him casting that early-fall day on Before Pool. With his new companion Rosalina keeping a watchful eye on his progress, Paul worked the pool. "I like muddlers in sizes 4 or 6," Paul said as he pumped out loops. "But lots of different flies will take fish. You'll see them surface for flies too."

Just as Paul Bugeja and Parker Wright mentioned, a variety of fly patterns will entice the landlocks. Streamers are very popular for sure, but plenty of salmon are taken on nymphs. I've found purple flies are effective for landlocks on spawning runs. Many years ago, on New York's Fall Creek, I took many landlocks running up from Cayuga Lake on big Hare's Ear–type nymphs. A size 10 Atherton Number 2 nymph accounted for many successful days, as did a lighter shade Atherton Number 1. Classic steelhead patterns also entice landlocked salmon.

A variety of fly patterns, from small nymphs to larger tandem streamers, entice Lake Memphremagog landlocked salmon. Scott Biron, a talented streamer fly tier, crafted the Sandbar Smelt Tandem (left) and Montreal White Belly Tandem (right). Other tandem streamers such as Champlain Special Tandem and Willoughby Rainbow Tandem are also effective.

Purple Matuka *Tied by Dave McNeese*

Purple streamers are very effective for enticing landlocked salmon. Purple dubbing and dyed purple guinea, along with fine silver tinsel ribbing, creates this colorful fly. It can be tied on smaller size hooks without jungle cock cheeks.

Orphan *Tied by Dave McNeese*

Drab shaded flies are also effective at certain times when targeting landlocks. This Orphan version was influenced by Myron Sprague. The body is created with olive-brown angora. The wing is brown bucktail topped with gray squirrel tail. A small gold tag adds a bit of flash.

Atherton Number 1 *Tied by Mike Valla*

Atherton Number 2 *Tied by Mike Valla*

UPPER CLYDE RIVER—ISLAND POND TO CLYDE POND

The Clyde River begins its 40-mile journey at the town of Island Pond as an outlet of Island Pond. The river flows past a mid-1800s hotel, then soon picks up cooler water from its Pherrins River tributary, the marvelous brook trout fishery described on pages 143–46. Adventurist fly anglers could launch at the outlet, float underneath the old hotel, then embark on a float/fish journey that will take perhaps 2 to 3 hours to reach a formal VF&WD access takeout downstream at Five Mile Square Road crossing. A section of Class II rapids exists between the Island Pond outlet and Five Mile Square Road.

However, the better option might be to sample the river downstream from the Five Mile Road access point, along the river that now completely changes its character. The river changes from tree-covered stretches, often clogged with woody obstructions and faster water, to slow-moving, wide-open boggy areas. I have not fished the reaches immediately downstream from the bridge crossing, so I turned again to VF&WD fisheries biologist Jud Kratzer to gain better understanding of the river from the Island Pond outlet and sections downstream. In response, Jud provided a complete overview of the Clyde River, which he called "a very interesting system."

Jud explained that the outlet from Island Pond is mostly warm water during summer, but brook trout no doubt pass through there at times. Once the Pherrins River dumps in its cooler water, downstream approximately 0.5 mile from Five Mile Square Road, anglers who fish from a kayak or canoe might get into some nice brook trout.

"This stretch of the river holds some of Vermont's largest stream-dwelling brook trout," Jud said, "all wild. There are a few reasons why this stretch supports such large brook trout. It is deep with lots of cover. It is very productive, and the water is cold. It is also extremely difficult to access and fish. Most of it is surrounded by nearly impenetrable swampy forest, and there are very few places that can be waded. The only way to fish most of it is by canoe or kayak."

Mileage downstream as it approaches the VF&WD launch put-in at Ten Mile Square Road opens up significantly as it flows through a warmwater fishery.

Once you get well downstream from Five Mile Square Road, beyond the trout-inhabited zone and all the way to Pensioner Pond, you'll be casting your warmwater species flies for pickerel, rock bass, fallfish, and perch. Jud said the cascades between Pensioner and West Charleston Ponds might hold a few trout that come down from above. The section downstream from West Charleston Pond Dam is worth trying for trout and landlocked salmon. "There are a few wild brown trout in this reach, especially near the dam," he said. "Depending on how many salmon get moved upstream of Clyde Pond, it could be worth fishing for salmon in the fall. Again, best chances are probably below the West Charleston Dam. There is no dam on Salem Lake."

Clyde River's section between Five Mile Square Road and the next VF&WD launch area at Ten Mile Square Road flows through wide-open areas that are inaccessible without a canoe or kayak.

Fly anglers often sample Clyde River stretches that have good structure and runs around Derby and Charleston, located off VT 105/VT 5A, upstream from the river's confluence with Clyde Pond. Take a view of the river off bridge crossings in this area. Clyde River Park, a public park in Derby, is popular with both anglers and paddlers. The small park, located on 11 Bridge Street, has parking next to picnic tables. You'll need the usual assortment of flies that you would cast on similar-looking water at the time of year you're visiting the area. Have a supply of stonefly nymphs on hand.

SPECIAL REGULATIONS

Note that there are the Special Regulations attached to the Clyde River. Make sure you review the official VF&WD Fishing Guide & Regulations syllabus available online and in print form. A variety of Clyde River regulations are in effect, ranging from C&R restrictions to stretches entirely closed to fishing. The syllabus describes all Clyde River regulations in an easy-to-understand format. September 1 to October 31 is the when you'll want to start scouting fishing reports.

White River a short distance upstream from the West Hartford Bridge, off White Lane Road near the Mill Brook tributary. The White flows through ledge rock areas that create fast water perfect for rainbow trout.
VALERIE VALLA

Connecticut River Basin

Waters that drain the Connecticut River Basin pass through some of the most beautiful areas in all of Vermont. Beginning in its northernmost region—the Northeast Kingdom—rivers such as the Nulhegan, Paul Stream, Passumpsic, and smaller brooks flow gently through areas flanked by mountains so breathtaking with fall foliage colors that it's sometimes difficult to focus on fishing itself. Forest Lake is located even more north in the Northeast Kingdom, near the Canadian border. Farther south, Vermont's Connecticut River shoreline accepts rivers that might be difficult to phonate—Ompompanoosuc and Ottauquechee—yet provide good fishing for wild brook trout in its headwaters to rainbow trout along their main stems.

The White River alone, from its birthplace as a trickle in Green Mountain National Forest to its expansive, wide-open mouth at the Connecticut in White River Junction, could easily occupy a fly fisher's time for an entire season. Then there's the mighty Deerfield River, also born in the Green Mountains, that provides fishing variety that in some ways surpasses all other Vermont waters. Its upper branches enter a series of reservoirs and tailraces that continue all the way into Massachusetts before the river finally sends water to the Connecticut.

Forest Lake/Averill Lakes—"Quimby Country"

Forest Lake, located a little over 20 miles north of Island Pond, is a shallow 62-acre trout pond that drains southeast into the Connecticut River. On your way to Forest Lake, you would have driven past the Pherrins River along VT 114 and Norton Pond. Forest Lake Road is off VT 114 just short of 5 miles north of the VT 114/VT 147 intersection in Norton. Take Forest Lake Road to Quimby Country, marked by signage. For convenience, Great Averill and Little Averill Lakes will be grouped with Forest Lake in this basin because of their close proximity, although the two are technically listed as members of the Memphremagog drainage system, as is true with the Pherrins River, Norton Pond, and Coaticook River. The Averill Lakes are located off VT 114, a relatively quick route to Canann and the upper Connecticut River.

Forest Lake's meager maximum 18-foot depth (VF&WD references state its depth as 13 feet) relies on spring seepages to support rainbow trout stocked annually by VF&WD. Brown trout also inhabit the pond. Over many years, Vermont has stocked varied trout species in the pond. Historical data shows that from the mid-1970s until about 1980, somewhere between 3,500 and 4,000 brook trout were stocked in the pond annually. In 1977, 11,000 brook trout fingerings were stocked. By 1981 the number of brook trout stocked dropped to 1,500. From 1979 until 1982, somewhere between 900 and 2,000 brown trout were stocked. Beginning in 1982, biologists continued to adjust best management practices for the pond and stocked only brown trout up until 2020. By 2020, only rainbows were stocked, and in 2024, 500 11-inch rainbows and 100 15-plus-inch 2-year-old rainbows were allocated.

Back in the early 1980s, then-Vermont fisheries biologist Leonard Gerardi reported reasons for discontinuing heavy brook trout stockings. Morphological and temperature

Forest Lake, high in the Northeast Kingdom near Canada, is synonymous with Quimby Country, the historic assemblage of cottages and great lodge that served as a relaxation destination for more than a century.

characteristics of Forest Lake were not favorable for brook trout. "Brook trout survival through the summer and following winter is apparently limited," Gerardi wrote in his 1983 report. That said, Quimby Country owner Gene Devlin told me during my stay at the facility that you might possibly still run into a brook trout and a big brown trout while fishing the small pond. But for all practical purposes, you'll be targeting stocked rainbows.

While Quimby Country owns the entire shoreline, public access to the lake is required by law, something Quimby's guest manual makes clear. A small unimproved public launch is located off Albert's Way, the gravel road that leads from Forest Lake Road to Quimby's. It's just beyond the Quimby Country signage. However, I advise booking Quimby Country and staying in one of its interesting cabins, alone or with fishing friends. Quimby Country isn't so much about Forest Lake itself. The facility is a good staging location as you make your way around the region and sample other surrounding lakes, ponds, and trout streams.

Signage along Forest Lake Road directs visitors to Quimby Country, a private historic fly-fishing and now general recreational and relaxation establishment. Quimby Country can serve as a great staging area for anglers in need of lodging in that remote part of the Northeast Kingdom.

A Quimby Country experience is well worth the cost, although most of the peak summer months require longer stays. Short-stay reservations are available during June and again in September and into October, some of the best fly-fishing periods in the area; check the particulars online. The main lodge, clubhouse, and 19 lakefront cottages are the only buildings around the lake. The

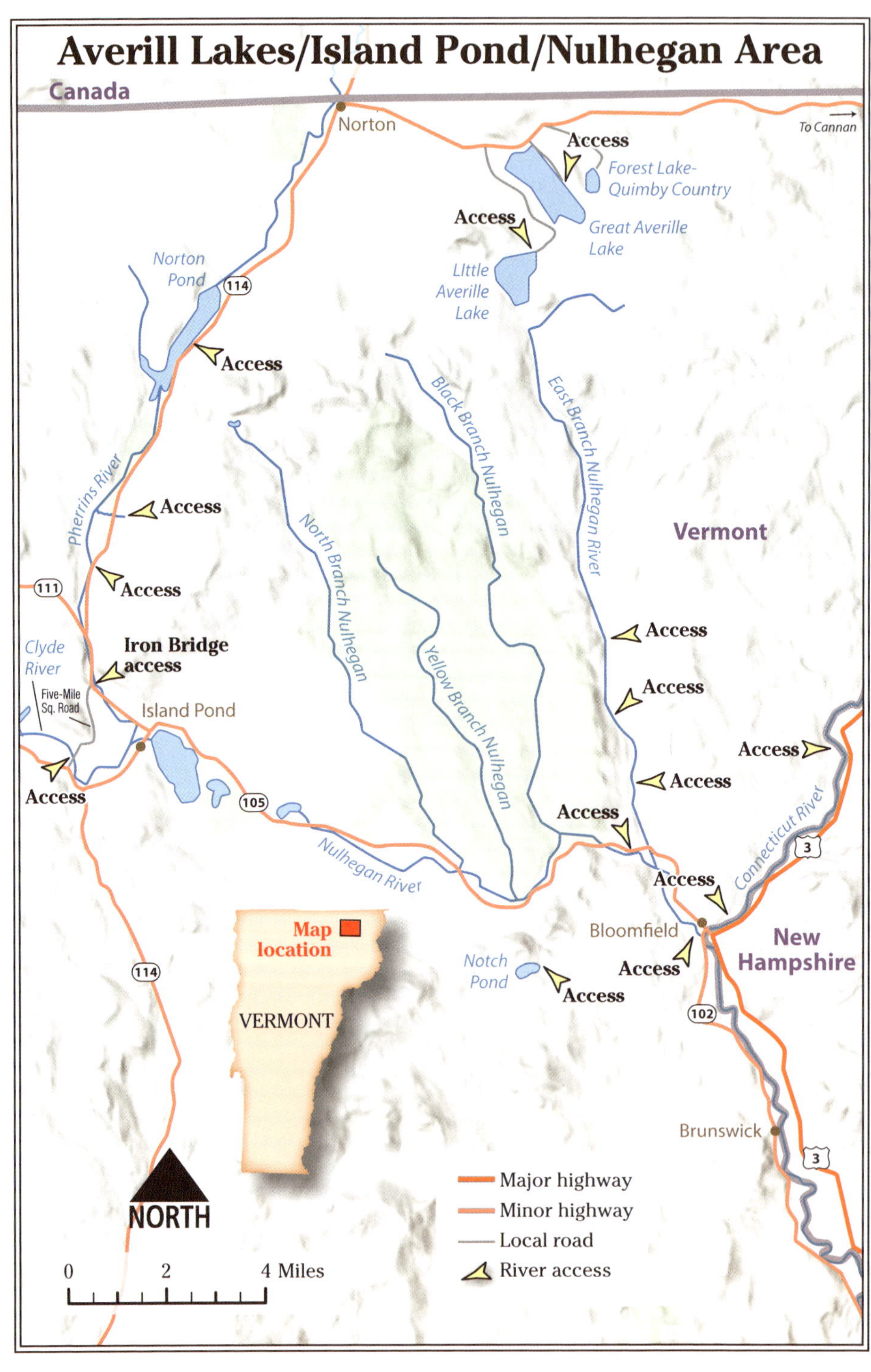

Averill Lakes/Island Pond/Nulhegan Area
Canada
Norton
To Cannan
Access
Forest Lake-Quimby Country
Access
Great Averille Lake
Little Averille Lake
Norton Pond
114
Access
Black Branch Nulhegan
East Branch Nulhegan River
Pherrins River
Access
North Branch Nulhegan
Vermont
Access
111
Yellow Branch Nulhegan
Access
Iron Bridge access
Clyde River
Access
Five-Mile Sq. Road
Island Pond
Access
Access
Access
105
Access
Connecticut River
Nulhegan River
3
Access
Access
Map location
Bloomfield
New Hampshire
Notch Pond
Access
Access
114
VERMONT
102
Brunswick
3
Major highway
Minor highway
Local road
River access
NORTH
0
2
4 Miles

Hare's Ear, one of many Forest Lake cottages named for classic trout flies. Hare's Ear is one of the smaller cottages. It's quaint wood-burning stove keeps occupants warm during the chilly seasons.

cottages range in size and amenities. All have full baths; some have small kitchens. The individual cottages are named for classic trout flies: Ginger Quill, Black Gnat, Greenwell's Glory, Parm Belle, Wickham's Fancy, and Jock Scott, to name a few. Valerie and I stayed in Hare's Ear, a small cottage, on a late September trip. On another trip we stayed at the larger two-bedroom Parm Belle cottage. The cottages are equipped with quaint wood-burning stoves that kept us warm at night during the cool late fall nights.

Besides the apt cottage names and fly-fishing opportunities, Quimby Country's history is also an attraction. In 1893 Charles Quimby became half-owner of a fishing camp on what was then named Leach Pond. By 1904 he owned the entire place. It was passed down to his daughter, Hortense, who renamed it Quimby's Cold Spring Club and expanded its mission to a relaxation destination not solely a fishing camp. After Hortense passed, others kept the club open for decades under its new name, Quimby Country. The Devlin family acquired full ownership of the lake and its 1,000 surrounding acres in 2018. The main lodge itself will transport you back into fly-fishing eras now past: a large mounted lake trout on the wall, framed classic fly prints, and an expansive library and sitting area where you can share stories about the day's catch.

Today, Quimby Country is a family recreational and relaxation destination. All-inclusive family vacation weeks are unique and include lots of activities. Fly fishing is definitely a focus activity. Numerous boats and kayaks are available across from the main lodge at no extra charge to guests. Owner Gene, a fly angler himself, can provide updated conditions. You'll likely find him fishing on Forest Lake during a *Hexagenia* hatch that arrives usually by the first week of July.

GREAT AVERILL LAKE

Quimby Country generally opens by mid-June and remains operational until about mid-October. By the time the facility opens, water temperatures are still favorable for lake trout on Great Averill Lake (aka Big Averill Lake or Great Averill Pond). Fly-fishing for lake trout that hold near the water surface, based on temperatures, is another reason to book Quimby Country, although you can also access the water via a VF&WD launch off Cottage

While access to Great Averill Lake, an 832-acre water with a maximum depth of 105 feet, is also available at a VF&WD public boat launch off Cottage Road, easy entry to the water is through Quimby Country lands via a carriage trail that leads from the lodge parking area.

Road on the eastern shore. On most oligotrophic lakes, lake trout head for the depths once surface water temperatures reach the high 46°F degree point. An easy quarter-mile trek down a carriage trail leads from the parking area next to the main lodge to Great Averill Lake. "The Rock" landmark gathering area for visitors is located at the shoreline. You'll find floating craft available for use at The Rock. While Great Averill Lake is primarily known for its abundant lake trout, stocked rainbows also inhabit the lake. Up until 2019, brook trout were also stocked. The official 2024 Vermont Fishing Guide Regulations list brook trout as having a "presence" in the lake.

If you're hauling your own boat, head for the public launch located off Cottage Road. If coming into the area from Island Pond, take Lake View Road off VT 114 to Cottage Road. The launch is about 0.3 mile from the Lake View Road/Cottage Road intersection. There is a VF&WD hanging sign along Cottage Road. (Lake View Road and Cottage Road are just around the corner from Quimby Country.)

LITTLE AVERILL LAKE

Like Great Averill Lake, Little Averill Lake is technically part of the Memphremagog drainage. As noted above, it's included here because of its close association with Forest Lake and gateway access to the Upper Connecticut River via VT 114. (The Upper Connecticut River is discussed in the next section.)

Little Averill Lake is easily accessed at the public boat launch off Jackson Road. The Jackson Road/VT 114 intersection is located about 20 miles north of the VT 114/VT 105 intersection in Island Pond. Take Jackson Road about 3.1 miles to a Y at the intersection with East Branch Road. Bear right and continue 0.2 mile to "Little Averill Boat Launch" signage. It's quite easy to launch your canoe or kayak. Valerie and I floated our lightweight

Little Averill Lake, a 470-acre lake high in the Northeast Kingdom, at the boat launch off Jackson Road. Little Averill Lake connects to Great Averill Lake via Averill Creek. Averill Creek delivers water from both lakes to the Coaticook River then quickly into Canada.

Hornbeck with little issue. The official 2024 Vermont Fishing Guide Regulations list the lake (more of a pond) as primarily a lake trout water, with a "presence" of brook trout. Brook trout were last stocked in the water in 2016.

Upper Connecticut River

While visiting and fishing Island Pond or Quimby Country area waters, you'll also want to fish the Upper Connecticut River, arguably one of the best trout destinations in the Northeast. While Vermont residents who hold fishing licenses are authorized to fish the entire Connecticut River, nonresidents will need to secure a New Hampshire nonresident fishing license to fish the main stem. A short-term license is reasonably priced. Be sure to review and understand all the special regulations that apply to Upper Connecticut River waters in both the Vermont and New Hampshire areas and portions of their tributaries.

Upper Connecticut Special Regulation reaches, the most popular mileage in both Vermont and New Hampshire, are within reasonable driving distances from Island Pond or Quimby Country area lakes and streams. While float- and drift-fishing are the most effective, wade-fishing can be enjoyed on many stretches along the Upper Connecticut River. If you're targeting trout along the uppermost river sections near Canaan, it makes perfect sense to wander upstream into New Hampshire and experience its fabulous brown, rainbow, and brook trout and landlocked salmon water.

The Vermont–New Hampshire border is only 4 miles upstream from Canaan. Pittsburg, New Hampshire, headquarters for fly anglers sampling the Upper Connecticut, is only another 5 miles from the border crossing. Brief mention of New Hampshire's extraordinary

Vermonter Dean Wheeler working the Upper Connecticut River a short distance upstream from the Nulhegan River's confluence in Bloomfield. Peak foliage season in early October and during lower water conditions are pleasant times to fish the Connecticut River.

Upper Connecticut River is discussed first, before delving into areas downstream into reaches shared by both states, beginning with the Canaan area.

NEW HAMPSHIRE'S UPPER CONNECTICUT WATER

The mighty Connecticut River that flows some 400 miles through four states is born at Fourth Connecticut Lake, a small shallow barely 2-acre pond next to the Canadian border yet still within the boundaries of Pittsburg, New Hampshire. One of four lakes called the Connecticut Lakes, isolated Fourth Connecticut Lake connects with a succession of three other lakes that are located progressively south—Third Connecticut Lake, Second Connecticut Lake, and finally First Connecticut Lake. Lake Francis, a man-made reservoir impounded by Murphy Dam, lies to the south of the four lakes.

The lakes are important to mention because they play an important role in the Upper Connecticut River fishery in New Hampshire's expansive Pittsburg region. Special Regulations boundaries are associated with the lakes. As far as Vermont is concerned, Murphy Dam and its discharge that flows to Canaan have special relevance. The New Hampshire Fish and Game Department's official fishing regulations define the following regulations:

Dam at Second Connecticut Lake to upstream side of logging bridge on Magalloway Road: This section, the most restrictive, limits the water to fly fishing only, catch-and-release for all fish species.

Magalloway Road bridge to inlet on Green's Point on First Connecticut Lake: Fly fishing only; daily limit two trout.

First Connecticut Lake Dam to the signs on Lake Francis: Fly fishing only, daily limit two trout; minimum length 12-inches. Most refer to this reach as the Trophy Section.

Murphy Dam at Lake Francis to New Hampshire/Vermont border: January 1 to March 31, catch-and-release only, single barbless hook, flies/lures; April 1–October 15, all legal methods. Daily limit 5 trout or 5 pounds

Anglers who intend to sample both Vermont's expansive Upper Connecticut River mileage as well as reaches that flow entirely through New Hampshire should secure a knowledgeable guide who knows the river well along stretches that flow through both states. While there are several Upper Connecticut River guides who can provide good information and instruction, one of the best is Nick Proulx, who runs both trips and instructional events out of Pittsburg. Nick, a former career fireman for 17 years, decided to leave the fire service and pursue guiding beginning in 2017. He calls his outfit Northern Water Guide Service; he also guides for Tall Timber. Northern Water Guide Service runs a couple of three-day fly-fishing schools, one in June and one in August.

Talk with Nick for 5 minutes and you'll quickly realize that he knows the Upper Connecticut in both New Hampshire and Vermont. Your chat will no doubt go beyond a few minutes; he'll cover a lot of ground with you prior to getting you on the water floating, drifting, or wading. He floats quite far down river through Vermont's best water too.

"The first few drifts we do are from downtown Pittsburg to the top section of the dam in Stewartstown. These two sections are slightly faster water with a lot of pocketwater fishing and are typically accessed with a raft. From the bottom section of the dam in Stewartstown all the way down to Lyman Falls, the majority of the river is relatively calm and slow, with faster runs divided up throughout the day. Most of this scenery is farmland cornfields and meandering fields with a river cut through the middle of it. On any given day we have the opportunity to dial in on multiple skills—dry flies, nymphing, and streamers, depending on weather and water conditions. We run both a Stealthcraft inflatable raft and, along with that, a 16-foot fiberglass drift boat."

Ask Connecticut River fly-fishing regulars what the best months to fish the river are and, not surprisingly, you'll get a range of responses. Some like fall fishing, a great time of year when water levels often provide for easy wade-fishing. Others suggest getting on the water when landlocks can be netted on New Hampshire reaches. Nick, who guides anglers with a range of abilities, suggests middle-of-the-season time slots.

"If I get an inquiry from someone about when to come up, I usually recommend June and July," Nick said. "There's plenty of bugs on the water, and conditions are more comfortable than the colder months for those who might be new to fly fishing. But that all depends on expectations—what fish species are in mind to target among them."

Nick pointed out that earlier months can be interesting. "We get a brief landlocked salmon run around Mother's Day, coming up the Magalloway Road area. Some come up for that experience, then we transition into fishing the trophy section. We have an abundance of caddis, along with Blue-Wing Olives pretty much throughout the season. A few March Browns and drakes here and there too. Landlocked salmon during late season typically come in the second or third week in September through to the close of the season on October 15." As far as landlocks coming out of Lake Francis, they're typically 18–20 inches, maybe a 22-inch fish once in a while, Nick said.

CANAAN, VERMONT, UPSTREAM TO CONNECTICUT RIVER STREAMBANK ACCESS, NEW HAMPSHIRE

If you're after a quick jaunt into New Hampshire while fishing the Upper Connecticut River along Vermont's sections, try the Connecticut River Streambank access. Approaching the Pittsburg area from Canaan, Vermont, head north from the VT 114/VT 253 intersection toward Beecher Falls. As you drive toward Pittsburg, the river flows flat but eventually picks up more structure as you continue north. At 1.8 miles north of Canaan, VT 253 turns left; instead, keep straight on River Road and continue following the river. You'll quickly pass flat water with road shoulder pull-off areas. You'll pass residential areas with little stream access on that side of the river until you cross the Vermont–New Hampshire border. Keep your eyes open for brief yet structured stretches where access is available along the road shoulder. Water levels will dictate whether wade-fishing is possible.

About a mile or so north of the Vermont–New Hampshire state line, you'll hit US 3. Turn left on US 3 and head into the Pittsburg area. After a brief access point, the river is hidden from view, flowing quite far from the road. At 2.8 miles north of the River Road/US 3 intersection, you'll encounter a formal fishing access point, the Connecticut River Streambank Access, marked by large signage and a small kiosk along the road shoulder, where you'll park. Access is made available through the courtesy of landowners, so please respect the privilege. Pick up any trash you might see as you make your way toward the river. A wide, usually maintained path will deliver you to the river, which has been out of sight.

Once you arrive at the tree line, you'll see the river flowing below from a vantage point above the stream bank. A crude footpath leads upstream and downstream. If water levels allow, you'll have to carefully locate manageable areas where you can get down otherwise sometimes steep banks along the water. It's a very nice scenic stretch, something

The Connecticut River Streambank Access in New Hampshire is located about 8 miles north of Canaan, Vermont. If fishing the Canaan, Vermont, reach, it makes sense to try a few Connecticut River access points along Special Regulations sections in New Hampshire.

A crude footpath follows the Connecticut River Streambank Access reach. Look for areas where entry into the water is safe and manageable.

you'll continue to encounter if you make your way farther upstream into New Hampshire's stretches.

CANAAN, VERMONT, DOWNSTREAM TO BLOOMFIELD

Canaan, Vermont, is a crossroads town located in the northeasternmost corner of Vermont. Sharing a land border with New Hampshire, Canaan touches the southwestern corner of the town of Pittsford in New Hampshire. Tourism, including fishing, is an important part of the town's economy. If you're lodging at Quimby Country, you can be on Upper Connecticut River water in no time. From the Forest Lake Road/VT 114 intersection that leads to Quimby Country, drive about 9 miles east on VT 114 to Canaan. Make certain you have the required fishing licenses before you throw loops on the Upper Connecticut River. It can be confusing, but here's how it's described in the official Vermont Fishing Regulations:

> All New Hampshire and Vermont resident fishing licenses are valid for the taking of fish from the Connecticut River [as defined for that portion of the river between New Hampshire and Vermont. Connecticut River means all waters of the river, including the bays, setbacks, and tributaries, only to the first highway bridge crossing said tributaries on the Vermont and New Hampshire sides]. All other nonresidents with a New Hampshire nonresident fishing license shall only take fish east of the Vermont low-water mark while on Connecticut River.

Boiling this all down, if you have a resident license from either Vermont or New Hampshire, you're all set to fish the entire river main stem. If you have a nonresident Vermont fishing license, you'll also need a nonresident New Hampshire license to fish the main stem.

The Upper Connecticut River Canaan, Vermont, bridge that crosses into West Stewartstown, New Hampshire. The Patriot Hydroelectric Station is on the left in the photo.

The Canaan reach is part of Special Regulations mileage that reaches from Murphy Dam in Pittsburg, New Hampshire, downstream to the bridge crossing into West Stewartstown into Canaan. Murphy Dam is located about 10 miles north of Canaan. From January 1 to March 31, the stream section is catch-and-release, barbless flies and lures only. From April 1 until October 15, all legal angling methods are allowed. As previously noted, more-restrictive fly fishing–only Special Regulation sections are also in effect in New Hampshire upstream from Murphy Dam, including New Hampshire's trophy section.

Access is available in Canaan on both sides of the river. Some fly anglers prefer accessing the stretch above the bridge crossing along the run downstream from the Patriot Hydroelectric Station. Patriot Hydro provides free public access to the river off Powerhouse Road. A couple of picnic tables and an information kiosk are located on Powerhouse Road a short distance from the bridge crossing. Another popular access point is available on the east side of the river in New Hampshire. Cross the bridge into West Stewartstown, New Hampshire. Solomon's One Stop Shopping store is located on the upstream side, immediately after crossing the bridge. Park in the rear of the store; a path will get you down into the river. Local conventional tackle anglers are frequently parked at the informal access point.

On one visit to Canaan, Valerie ran into a very focused local town angler who was casting from the large parking and launch area on the downstream Vermont side of the bridge. Valerie engaged him in discussion about fish species he had latched into in that section. While still casting—and intently focused on the river, not us—he said, "Well, I've caught rainbow trout, brown trout, and even a landlocked salmon right here. And suckers." (By the way, he was all of 11 or 12 years old.)

COLEBROOK ACCESS

Another popular access area is located at a bridge crossing about 8 miles south of the VT 103/VT 253 intersection in Canaan, off VT 102 and about 13 miles north of Bloomfield. At the bridge, cross over into New Hampshire, then take an almost immediate right onto Bill Bromage Drive. You'll see a very large open area that leads to the river, a popular launch site for recreational canoers and kayakers following portions of the Northern Forest Canoe Trail and Connecticut River Paddlers' Trail. A kiosk with maps is located at the far end of the lot. Water levels, of course, will dictate if you can safely wade-fish the nice run downstream from the bridge.

From Colebrook, the river flows through sections along both dairy farms and woodlands. A 14-mile journey delivers floaters to access points at Bloomfield and the mouth of the Nulhegan River, passing the Columbia Covered Bridge and the runs that traverse along Lyman Falls State Park.

The run downstream from the bridge crossing into Colebrook, New Hampshire, as it appears during lower water conditions. Access is easy at a popular Northern Forest Canoe Trail launch site next to the bridge. Wade-fishing is easy when water levels allow.

LYMAN FALLS STATE PARK

Sometimes difficult to locate off VT 102 due to missing signage, Lyman Falls State Park is a nice area when water levels provide safe wading. There's not much to the park aside from trails, a couple of picnic tables, firepits, a privy, and an information kiosk next to the river. Fly-fishing floaters often use the New Hampshire side of the river as a takeout point.

The run at the breached dam site at Lyman Falls State Park.

Recreational floaters often use the site as a stopping point or camping point for a longer journey down the Connecticut River.

The park entrance is across from a very large barn structure on VT 102. If driving south from Canaan and the Colebrook reach, it's located just after Mill Brook Road. If approaching from Bloomfield, the entrance is just a couple hundred feet before you reach Mill Brook Road. Pull into the parking area that's blocked off by large rocks. A pleasant yet long walk down a footpath through woodlands leads to the river.

A Special Regulations section exists from a point 250 feet below Lyman Falls Dam in North Stratford, New Hampshire, downstream to a point 1,600 feet upstream from the bridge crossing in North Stratford. (The bridge crosses the river into Bloomfield, Vermont.) This river stretch is catch-and-release, barbless flies and lures only.

BLOOMFIELD

A couple of good wade areas are available off VT 102 in Bloomfield. One is located at the Nulhegan River's confluence with the Connecticut River. Another nice access is located 0.1 mile off VT 102 north of the VT 102/VT 105 intersection in Bloomfield. You'll be fishing the barbless flies and lures–only, catch-and-release section that leads upstream to Lyman Falls.

Park along the road shoulder at a point where you'll see the beginning of a VT 102 guardrail. Some anglers park in a grassy area just off the road. Keep in mind that this is private property, although it's not posted. Protect continued access by picking up any trash

Looking downstream at the confluence of the Nulhegan and Connecticut Rivers from the VT 102 bridge in Bloomfield.

The Upper Connecticut River supports very large brown trout in both Vermont and New Hampshire. NICK PROULX

you might encounter as you make your way to the river. A crude footpath from the road winds and turns quite a distance before you finally arrive at the river.

When water levels are lower, such as times in late season, wade-fishing is easy and manageable. I ran into Vermonter Dean Wheeler one early October day while he was gearing up to work the long pool where the footpath joins the river. “It’s a nice time of year,” Dean said as he laid out his fly rod on his vehicle. “I purposely headed here after checking USGS water levels.” Dean has caught rainbows and brown and brook trout along this reach. “It seems that whenever I get rainbow trout, they’re all the same size,” he said.

There’s no lack of rainbow, brown, or brook trout along the Upper Connecticut River. Substantial numbers are stocked from Pittsburg way downstream into Vermont areas from Colebrook through Bloomfield and farther south into Stratford. First and Second Connecticut Lakes are stocked with landlocked salmon.

Nulhegan River

The 15.1-mile-long Nulhegan River, a Connecticut River tributary, is tucked into one of the most beautiful areas not only in Vermont's Northeast Kingdom but all of the Northeast. The Nulhegan River and its branches flow out of the spectacular Nulhegan River Basin. While sampling the Nulhegan system, you'll want to stop at Silvio O. Conte National Fish and Wildlife Refuge Center located off VT 105, about 10 miles east of Island Pond. A major attraction during the colorful fall foliage period from late September into October, the view of the basin area that creates the Nulhegan branches is a must-see.

Moose are frequently sighted along the Nulhegan; its boreal wetlands provide perfect habitat. Abenaki inhabitants from back in time called the Nulhegan River the "Deadfall Trap River" in their language. "Long River"—the Connecticut River, in their vernacular—accepted the Nulhegan in what is now the Bloomfield, Vermont, area. The river system is protected by the Nulhegan Basin Division of the Silvio O. Conte National Fish and Wildlife Refuge.

Logging operations, with their camps, settled the area during the nineteenth and twentieth centuries. Log drivers perched precariously on rafts of logs floated the Nulhegan and surrounding streams such as Paul Stream, also a Connecticut River tributary. Those log raft drives have been replaced by recreational paddling—and fishing. However, logging activity had negative effects on the river's fisheries, something that's being studied and identified to this day.

The Nulhegan River system can cause a degree of confusion for fly anglers visiting the area, given its roughly parallel four branches that join the main stem (see Averill Lakes/Island Pond/Nulhegan Map H Area map). The main stem is created as an outlet of Nulhegan Pond next to VT 105 in East Brighton (just east of the town of Island Pond).

The East Branch of Nulhegan River along East Branch Road, a couple of miles upstream from Bloomfield. The East Branch joins the main stem 2.3 miles before the river dumps into the Connecticut River at Bloomfield. It's one of four branches that combine their waters before joining the Connecticut. Strategic Wood Additions habitat enhancements were done about 8 miles upstream from this location. VALERIE VALLA

View of Nulhegan Basin from Silvio O. Conte National Fish and Wildlife Refuge visitor center. Nulhegan River's branches are born in one of the most spectacular areas of the Northeast Kingdom.

In an order of west to east, the first branch in line is the roughly 15-mile-long North Branch that arises as an outlet of Unknown Pond in Averys Gore, not far from Lewis Pond, situated to its east. Biologists report abundant wild brook trout upstream from Four Mile Road bridge, but it's a bit cramped for fly fishing. Then comes the Yellow Branch, also about 15 miles long, which meanders largely through a boggy area. Black Branch is a longer stream; its nearly 30-mile length begins on the slope of Round Mountain then grazes the slopes of Potash Mountain and Little Potash Mountain ridgelines. Biologists report abundant wild trout upstream of Yellow Bogs, also cramped conditions for fly fishing. Nearer to Bloomfield than the previous branches, it enters the main stem off the slopes of French Mountain, close to VT 105.

Finally there's the East Branch, which arises well to the north near Little Averill Pond, tucked into the extreme northeast corner of Vermont's Northeast Kingdom. East Branch, the closest branch to the Connecticut River, is strengthened by picking up Spaulding, Fisher, and Mink Brooks along its path. You'll cross the East Branch on VT 105 just a couple of miles from Bloomfield.

Access to the East Branch is available along East Branch Road. You'll notice the river falls in and out of view, but you'll have no problem getting into the water at a few plank bridge crossings that are located upstream. As you drive along East Branch Road, a gravel road, you'll encounter "mile marker" signs along the road shoulder. The first stretches that come into view are wide open. However, the stream changes in character in its upstream reaches.

You'll find a few areas that are likely brook trout water. A nice stretch is located 3.2 miles upstream from VT 105/East Branch Road, at the first plank bridge crossing. Continuing north on East Branch Road from the crossing, you'll come to another small bridge crossing in about 1.7 miles. Bypass this area. In less than 0.5 mile or so from the crossing, the stream comes into view and closely parallels the road. Parking and access are easy here, and you'll be able to quickly get into the water.

Another plank bridge crossing is located at Mile 8, a little over 3 miles upstream from the previous crossing. The stream here is very small water, but access is easy. You'll need

Plank Bridge crossing, located 3.2 miles upstream from VT 105/East Branch Road. The character and geomorphology of the stream is much different than found on the East Branch's lowermost reaches.

a short rod and hip or knee boots. On your way to Mile 8 and a third plank bridge crossing over the Brouillard Brook tributary, you'll notice the road is quite stony but well maintained. It's best to sample the uppermost East Branch reaches during late season, when the road is dry and drivable, even with a small sedan.

VF&WD fisheries biologist Jud Kratzer and his team did most of their Strategic Wood Additions habitat project at the area where Spaulding Brook joins the East Branch, a bit farther up the road from the Brouillard Brook crossing. Jud shared his opinion that some of the best brook trout fishing on the East Branch can be found it that section.

Having picked up all four branches, the Nulhegan River stages its final approach and entrance into the Connecticut River, just a couple hundred feet downstream from the VT 102 truss bridge crossing, a quick turn off VT 105 in Bloomfield Village. You'll easily locate a footpath that leads from a parking area next to the bridge to the river. The access

The Nulhegan River at the VT 102 bridge crossing in Bloomfield, located a quick turn off VT 105. The Nulhegan enters the Connecticut River just a couple hundred feet downstream from the bridge. Sizable caddis flights come off the river along this reach in mid-May. Parking and easy access are available at a small lot next to the bridge.

MV Fathead Minnow (top) and MV Emerald Shiner (bottom) *Tied by Mike Valla*

Fathead minnow tails are tied with olive cashmere goat hair. Lion's Brand Fishermen's Wool forms the body that's wrapped with silver tinsel. The wing is created by dying brown bucktail a yellowish shade. Blue Ice dub creates shimmer. Emerald Shiners are very effective for enticing rainbow trout, mostly in lakes but also on most any river. Peacock and Pink Ice Dub Shimmer provides flash.

is popular with paddlers, who can quickly drift down into the Connecticut River. An interesting Nulhegan information kiosk is located at the site. Fish upstream from the bridge along the faster water, where you might pick up a rainbow trout during spring or fall. I ran into a terrific Grannom caddis (*Brachycentrus*) swarm on May 20, 2023. You'll want to fish one of Tom Baltz's CDC Grannom Pupa flies during the emergence. Otherwise, try fishing small streamers that suggest small baitfish. Target the spunky rainbows during early spring. Make sure you read and understand the Special Regulations that apply to the Connecticut River, especially for nonresidents of Vermont. Vermont fishing regulations are available online and in print form.

The East Branch is set apart from the other branches because it receives 600–800 rainbow trout stockings by VF&WD from 6 miles upstream of the VT 105 bridge crossing downstream to the VT 105 bridge crossing. However, all four Nulhegan branches, along with other watershed streams such as Paul Stream, have been the focus of VF&WD's ongoing long-term efforts to restore trout habitat. Several years ago, VF&WD and other collaborators undertook Strategic Wood Additions habitat enhancements and restoration, a project that is still moving forward.

Rainbows can be had along Nulhegan's East Branch and the river's lowermost reach in Bloomfield. Small streamers that resemble baitfish work well.

Small streamers that suggest shiners and other baitfish work well on Nulhegan's reaches.

VF&WD fisheries biologist Jud Kratzer, who kindly forwarded his extensive research report findings, has led the charge after his studies pointed to lack of in-stream woody material as a primary culprit in limiting wild brook trout biomass. "Chop and Drop" practices are employed on smaller streams, felling trees into the stream and sometimes anchoring them with boulders and rocks. Collaborative efforts with Trout Unlimited, the US Fish and Wildlife Service, and the Weyerhaeuser Corporation have been ongoing since 2012.

Kudos to Jud and his team, who have devoted so much time and resources to this important project. The findings of Jud's 2023 Strategic Wood Additions report are encouraging: "Annual electrofishing surveys in the East Branch of Nulhegan River and its tributaries from 2012 to 2017 found that brook trout biomass tripled, on average, at treated sites after three years."

Notch Pond

Notch Pond is a lovely small brook trout pond located not far from the Nulhegan River. VF&WD is studying the pond. You might get a brook trout with a fin clip. You'll need a small floating craft to effectively fish the water and get into the brookies. Located in the West Mountain Wildlife Management Area, the pond is relatively isolated. You won't be able to drive to the pond itself; a 0.5-mile trek (all uphill, by the way) from the gravel

Located in the West Mountain Wildlife Management Area, very near Nulhegan River, Notch Pond and its shoreline are entirely public. A couple of camps are along the shore, grandfathered by leases. VALERIE VALLA

access road where you'll park will deliver you to the pond.

From Island Pond, drive east on VT 105 a dozen miles or so to Notch Pond Road, marked by a Vermont Fish & Wildlife West Mountain Wildlife Management sign. If you passed the Silvio O. Conte National Fish and Wildlife Refuge visitor center, you've gone 0.8 mile too far. If you're approaching from Bloomfield, driving west, Notch Pond Road is 7.2 miles from the VT 102/ VT 105 intersection.

Take gravel Notch Pond Road (a bumpy road in areas) 1 mile to a fork. Go right at the fork and continue another 1.1 miles to an

Notch Pond Trail follows Notch Pond's outlet entirely uphill. A lightweight Kevlar Hornbeck mounted to a metal-frame backpack makes the hike to the pond easier. VALERIE VALLA

Southern Redbelly Dace *Tied by Mike Valla*

A Southern Redbelly Dace bucktail is another good choice for pond brook trout.

area on the left side of the road where you can park. Notch Pond Trail (unmarked) is directly across the road. You'll see signage indicating that the trail is a private snowmobile trail used by a couple of camps located along the pond's shoreline. The camps, now on leased land, were grandfathered in years ago after the state took control of the area.

The trek up the mountain is about 0.5 mile long, all uphill. As you get closer to the pond, you'll come to a fork. It doesn't matter which direction to choose, as they both end up at camps, but pond access is not a problem. The alternate way into the pond is up at the power lines. It's flatter but also wet and still 0.5 mile in. The road going up to the power lines is pretty bad and bumpy. You're better off eating your Wheaties before trekking up the mountain.

Paul Stream

Historically a spawning tributary for Atlantic salmon heading up the Connecticut River, Paul Stream is known for its brook trout, from both natural spawn and stockings. At times you might be surprised to latch into a rainbow trout that wandered in from the Connecticut River. Accessed by driving up Maidstone State Park Road, off VT 102 south of Bloomfield in the Northeast Kingdom, the small tumbling stream has ample access. Beginning at the intersection of VT 102/Maidstone State Park Road, the gravel yet well-maintained road

Tucked away in the Northeast Kingdom, quaint Paul Stream, a small 10-mile-long tributary of the Connecticut River, provides brook trout experiences along its tumbling runs and pools. You might also encounter other species that run up into the stream from the Connecticut River, such as rainbow trout. VALERIE VALLA

Paul Stream during early spring, at the West Mountain Pond Road bridge crossing intersection with Paul Stream Road. The stream tumbles down below the bridge, but again maintains a moderate flow along nice runs and small pools.

makes its way to Paul Stream Road, on the right, in 2.2 miles. An information kiosk is located at the intersection of both roads.

While driving toward Paul Stream Road, you'll catch glimpses of the stream that fall in and out of view. You'll also notice colorful artistic craftwork nailed to trees spaced along the road—everything from butterflies to cardinals. An owner of a local getaway cabin once told me they just serve as welcome gestures. You'll find most of the local cabin owners along the stream to be helpful and knowledgeable.

It's possible to park along the road shoulder where the stream runs relatively close to the road and get into the water at such points. But access is much easier along Paul Smith Road. Note that Paul Smith Road is gated during early fishing season, but locals told me the gate is always open for vehicle use by Memorial Day.

If the road is gated, park at the kiosk and walk around the gate and up the road a couple hundred feet to get a view of the stream at a small bridge crossing on West Pond Mountain Road at the West Pond Mountain Road/Paul Stream Road intersection. The stream tumbles through rocks and boulders downstream from the bridge crossing. Paul Stream Road continues to the left at the split; normally that's the direction you'll take to get into the water, since the stream largely follows very close to the road.

One of the easiest access points along Paul Stream is located along Paul Stream Road, about 1.9 miles from the VT 102/Maidstone Park Road intersection. I call this the Gravel Pit Reach, since you'll park directly across from a gravel pit at a convenient cubby hole area directly next to the water. A short footpath gets you quickly into water. Fish upstream and downstream from the access point. A 7-foot, 4-weight rod will serve you well. Wild brookies, most quite small, hang out in the runs along small boulders and rocks. They'll hit small Elk Hair wing dry flies.

Like many others, this stream was highly impacted by logging and clear-cutting during the 1800s and into the twentieth century that altered habitat and impacted the fishery. Today, local cabin owners along the stream are quick to share with visitors the history of area logging camps and log drives down the stream itself back in that time. As a result of Strategic Wood Additions habitat improvement efforts by VF&WD biologist Jud Kratzer and collaborators, Paul Stream has already shown improvements in wild brook trout numbers upstream from Ferdinand Bog and has not been stocked since 2021. Wild brook trout numbers downstream of Ferdinand Bog still justify supplemental put-and-take stockings. VF&WD typically stocks 800 brook trout along the stream from Ferdinand Bog all the way to the mouth at Brunswick.

You'll want to check out Maidstone Lake at the state park, a perfect place to camp and fish while sampling not only Paul Stream but also other local trout waters, such as the Nulhegan River. More than 30 RV/tent sites and 25 lean-to sites are available, with amenities such as hot and cold running water and flush toilets. With a maximum 121 foot depth, Maidstone's nearly 800 acres are inhabited by rainbow trout, lake trout, bass, and other species. Kayaks, canoes, and rowboats can be rented at the park. Your own craft is welcome.

Moose River

The headwaters of Moose River, both its east and west branches, are located just a few miles directly west of Paul Stream's headwaters, near East Haven Mountain and the East Haven Range. After joining together several miles north of Gallup Mills, the Moose flows several miles south through that hamlet then continues along River Road and Victory Road about 10 miles to US 2 at North Concord.

Headwaters of Moose River, upstream from Gallup Mills. While the focus is usually on brown and rainbow trout stocked annually along reaches from Gallup Mills downstream to Concord Village, naturalized brook trout are found along upstream headwaters. VALERIE VALLA

Along its route south toward Victory, the river passes through boggy mileage not that appealing for fly fishing. A couple of stretches at Victory Basin Wildlife Management Area (marked by signage), about 2.6 miles downstream from Gallup Milles, are worthy of sampling. A footpath next to the road leads into the river. Once you get downriver 6 miles from Gallup Mills, the river gains structure that appears more like a classic trout stream. The bridge crossing at the corner of Victory and Willson Roads is particularly nice, structured water with riffles and runs. It's a good place to sample. VF&WD typically stocks about 800 rainbows and 1,100 browns during mid-May from Gallup Mills to Concord Village.

The Moose River continues flowing roughly along US 2 another dozen miles until it joins the Passumpsic River at the intersection of Concord Avenue and Elm Street. The Moose shares the Passumpsic's Special Regulations stretch in St. Johnsbury—water from its confluence upstream to the Concord Avenue bridge crossing. The short section is open to fishing only from the second Saturday in April until October 31. Large triploid rainbows are stocked in the Special Regulations section.

While fly anglers largely target the browns and rainbows around Concord and St. Johnsbury, brook trout can also be netted in the Moose River. "If you're after brook trout," John, a friendly local fly angler who owns a cabin near Gallup Mills told me, "get into the stream in its headwaters." John, who has fished the stream for many years, staging his activity out of the family cabin handed down to him, quickly grabbed a pen and sketched out where I should cast my line. "I've netted some typically small yet colorful brookies up there over the years." VF&WD biologist Jud Kratzer advised me that the reach upstream from Gallup Mills is considered "low abundance" brook trout water. But if you're after a degree of isolation while fishing along a portion of the Moose River removed from congestion, you'll want to give it a try.

I passed by the Moose near Gallup Mills a few times in the past but never considered the headwaters upstream. On one visit my wife, Valerie, suggested that we venture up "that dirt road that follows the stream into isolated areas." My response was, "I'm not driving up there—the road hasn't been graded yet this spring." It wasn't until much later, on subsequent trips, that I took John's advice. If you desire isolation and small-stream fishing, head for Radar Road, located off Victory Road in Gallup Mills, right near the bridge crossing

Bridge crossing at the corner of Victory and Willson Roads, about 6.5 miles downstream from Gallop Mills. The Moose River acquires good stream structure in this area.

The Moose River at its confluence with the Passumpsic River in St. Johnsbury. The short stretch from its junction with the Passumpsic upstream to the Concord Avenue bridge crossing is part of the Passumpsic's Special Regulations section. (The Passumpsic is shown in the photo flowing along the tree line.)

in the town. Locals will be more interested in telling you about a relic from the past than brook trout fishing possibilities.

Radar Road is probably better known for the 1950s Cold War–era North Concord Air Force Station radar site, located on the remote East Mountain ridgeline. Called "a creepy place," ruins of the old structure, abandoned in 1963, still remain. Adding to its creepy reputation is the old report of a UFO siting up there, spotted by the military shortly before the site's closing. Locals enjoy sharing folklore about a possible alien abduction during that time frame.

At a point 0.2 mile from Victory Road, you'll see an information kiosk that welcomes public recreational users to the conservation easement managed by the Vermont Agency of Natural Resources. The area is a working timber-producing forest that welcomes anglers and other users. Visitors are asked to give logging trucks the right-of-way and to respect closed roads. By the way, the old radar site is quite removed from the areas you'll be fishing—no fears of alien abductions by UFOs.

At first, as you drive along the dirt road, the river is largely out of sight until just over 2 miles, when it comes into view flowing next to the road. Stay straight as you drive along the stream. The road name changes. You can get into the water at that point with a bit of effort, but a better strategy is to continue 2.7 miles or so from the kiosk to a small bridge that crosses the Moose River, here more of a brook. Plenty of parking is available next to the bridge. Fish both upstream and downstream from the bridge. Use small bead-head nymphs or small dry flies. This is great wet-wading water in mid-June and through summer. I've had better luck fishing upstream with Ed Ostapczuk's Elk Hair Stone Fly in #14-16.

After fishing Moose River's headwaters, if you're up that way, it makes sense to take a quick drive west to the nearby East Branch of Passumpsic River. At the Radar Road/Victory Road intersection, drive west 7.8 miles to VT 114, the road that follows the East Branch.

Passumpsic River

The Passumpsic River is born from its east and west branches, which join just outside of Lyndonville at the southern end of Darling Hill after flowing several miles from their beginnings near New Haven and near the south end of Lake Willoughby, respectively. From Lyndonville, the now main stem snakes another 12 miles to St. Johnsbury, where it gathers the Moose River near Concord Avenue and Elm Street at a Special Regulations stretch. With its combined water volume, the Passumpsic continues past St. Johnsbury, pushing 10 miles through wide, flat water toward its confluence with the Connecticut River at Nine Island. Along the mileage between St. Johnsbury and the Connecticut River, the Passumpsic passes over three dams. The final dam is located about a mile upstream before the river enters the Connecticut.

Mother Nature hasn't been kind to the Northeast Kingdom, including the Passumpsic and other streams, in the last couple of years. Devastating flash floods in mid-July 2024 ripped through the Northeast Kingdom, almost to the day of the catastrophic flooding that occurred in Vermont in 2023. Lyndonville and other areas were hit hard, only to be stomach-punched again the end of July 2024. The July 30 event dumped 8 inches of rain in 6 hours in nearby St. Johnsbury, once again wiping out roads and homes along the river system. The storm has been called a "thousand-year rain event." Thousand-year rain events have only a 0.1 percent chance of a specific amount of rain falling in any given year. While rivers will require time to heal, if you're up that way, please patronize the businesses that depend on recreational visitors and tourists, including fly anglers. Please tread lightly on the headwaters.

Passumpsic River—East Branch

Passumpsics's East Branch begins near Hancock and Sukes Ponds, north of East Haven near VT 114, just west of Moose River's headwaters. As it makes its way south toward Lyndonville, the East Branch picks up a dendritic group of small tributaries. Near the Caledonia–Essex County line, water from Bean and Mill Brooks feeds the stream at East Haven. Dish Mill Brook, a major tributary that supports wild brook trout, enters the East Branch downstream in East Burke Village.

Keep in mind that VF&WD doesn't stock the East Branch, and most of the wild brook trout are more concentrated upstream in its headwaters. While I have picked up brook trout downriver, stretches upstream from the VT 114/Lost Nation Road intersection run cooler and contain more trout. I asked biologist Jud Kratzer if he could chime in with any survey data on East Branch brook trout populations. His remarks were quite interesting:

"We have a lot of data on the East Branch Passumpsic, less on the West Branch. The reason we have so much data on the East Branch is that we used to stock salmon fry there as part of the Connecticut River salmon restoration program. It was good salmon habitat because the temperature is marginal for trout from about Lost Nation Road (Mill Brook) downstream to the mouth. We sampled a station in East Haven and one at Burrington Bridge Road every year from the 1990s through about 2010, and we would catch a few brook trout every few years. The East Branch runs colder upstream of Lost Nation Road, so there are more brook trout. Our most recent sampling up there was in 2011, and we estimated about 678 brook trout per mile, which is not great, even by Northeast Kingdom standards."

The headwaters are quickly accessed if fishing the Radar Road reach along the Moose's headwaters. Take Victory Road west about 7.8 miles from the Radar Road/Victory Road intersection in Gallup Mills to VT 114, which will deliver you to the East Branch. From

East Branch of Passumpsic at the Burrington Bridge Road crossing off VT 114, just upstream from Lyndonville. This reach is a designated Conservation Land Access Point, part of the Passumpsic Valley Land Trust.

that point, you can sample stretches upstream then reverse course and head downstream, fishing reaches that lead to Lyndonville. If approaching from its confluence with the West Branch at Lyndonville, you'll drive up VT 115 to access points, mostly at bridge crossings.

From the VT 115/US 5 intersection in Lyndonville, you'll come to a bridge crossing at Lily Pond Road next to a recreation field. Try casting for a while along that reach before continuing upstream along VT 114 just short of a mile to the Burrington Bridge Road crossing. You'll notice a quaint covered bridge that parallels the road bridge. Park along the road shoulder, make your way into the river, and fish upstream from the bridge crossing. I use my 7-foot, 4-weight rod along this stretch, casting dry flies along the runs and riffles.

You'll cross the river again in less than a mile while continuing to drive up VT 114, but continue to White School Road, about 5 miles north of the VT 114/Burrington Bridge Road intersection. Drive down the road to a bridge crossing, where the stream is quite nice. Parking can be an issue here, but you can squeeze in near the designated Kingdom Bike Trail (please don't block the trail). In just 1 mile up the VT 114/White School Road intersection, you'll find a large formal parking area along VT 114 next to the river that provides easy access. (Just before reaching the large parking lot, you would have passed Victory Road, noted above as a route to the headwaters of the Moose River near Gallup Mills.)

Other easy access areas are located as you continue your way north on VT 114. Try the run located behind the East Haven clerk's office along Community Building Road, off School Street (1.6 miles north of the formal parking area). School Street, a circular road that intersects VT 114 twice, crosses the stream at two locations. You could get into the water at the lower bridge crossing on Community Building Road, fish upstream to the upper crossing, then walk back to the point of beginning. At about a mile upstream from the upper School Street/VT 114 intersection, you can get access at the River Road Crossing. (If parking next to the bridge, don't block the fire hydrant.)

While the East Branch gets quite small as you continue north along VT 114 upstream from the VT 114/Lost Nation Road intersection, keep your eye out for a few stretches flowing next to the road where fly fishing is still possible for another few miles upstream. You'll want a 6- to 7-foot rod while negotiating such small water.

Passumpsic River—West Branch

Beginning in the town of Westmore, east of Lake Willoughby near Mount Pisgah, the Passumpsic's West Branch flows south along VT 5A and US 5 toward Lyndonville and its confluence with the East Branch. "The West Branch runs colder and even supports a few trout at its confluence with the East Branch, but trout abundance is still low," VF&WD biologist Jud Kratzer told me.

Access is more apparent and readily available on the East Branch compared to the West Branch. One might assume for that reason alone that it's the better of the two branches. "An angler might draw that conclusion," Ted Benoit, owner of Lead & Tackle Co. outfitters in Lyndonville, told me during an enjoyable chat we had in his shop, located near the east and west branches. "The East Branch has pretty thin water," he said. "The West Branch displays some interesting features and structure along stretches that are somewhat out of sight."

Ted's shop serves as a gateway of sorts for anglers heading into many waters in the Northeast Kingdom. You'll want to stop in while fishing not only the Passumpsic system but also on your way to other lakes, ponds, and streams in the area. The shop carries flies tied by local fly tiers. "Stonefly Nymphs are very effective on the Passumpsic River," Ted said, dropping a couple of Al Pitt's Golden Stonefly nymphs in my hands. Al Pitt operates a

West Branch of Passumpsic at Bugbee Crossing Road, located a couple of miles south of West Burke off US 5. While brook trout are much more numerous well upstream in the headwaters, brook trout inhabit this stretch.

small fly shop out of his home in Lunenburg, Vermont, not far from the Connecticut River. Al's Maple Syrup nymph variation is also effective on the Passumpsic River.

West Branch access points are not as prevalent, and much of the stream flows well away from roads. However, a couple of sites are worthy of mention, and a cast or two. While not what I'd call prime fly-fishing water, the stretch that runs near the intersection of US 5/VT 5A in West Burke, 15 miles north of the US 5/I-91 intersection in Lyndonville, is worth a quick look. The Sutton River tributary, small water, joins the West Branch in West Burke. This is short-length rod fishing, where hip boots or summer wet-wading is possible. Tight parking next to the bridge crossing is possible next to the West Burke Fire Department building, but be careful not to block the hydrant.

Another access is located about 3 miles south of West Burke, at Bugbee Crossing Road off US 5. Park on the far side of the small concrete bridge, blocked off with large rocks. The stretch flows downstream under the railroad crossing, at a nice pool that's often difficult to access on high water. "The stretches downstream from the railroad crossing have some very nice pools that hold brook trout," Ted Benoit told me. It's much easier to access the water upstream from the bridge crossing. From the shoulder parking area, cut through the small clearing and head into the wooded area. Remnants of a footpath will lead to the stream. It's a pretty area that provides a degree of isolation. "Our most recent sampling was below Bugbee Crossing Road in 2015. We estimated about 617 brook trout per mile plus a few juvenile brown trout thrown in," Jud Kratzer noted.

Passumpsic River—Main Stem

Much of the fly-fishing activity on Passumpsic's main stem occurs right smack in the middle of St. Johnsbury, along the Special Regulations section that flows 2.5 miles from Arnold Falls Dam downstream to Cage Dam. While Vermont Fish and Wildlife also stocks several thousand rainbow and brown trout in sections running from the confluence of the east and west branches downriver to the Connecticut River, the Special Regulations stretch downstream to I-91 receives large 2-year-old browns and rainbows. The river is stocked annually sometime during May. A short section of the Moose River at its confluence with the Passumpsic is also included in the Special Regulations. Those sections are open from the second Saturday in April until October 31.

Access is easy at a location where the Moose River joins the river at the corner of Concord Avenue and Elm Street. The lot at Fred Mold Park, which features an overlook fishing platform on Concord Avenue, is small and can accommodate a few vehicles. It's much easier to park along Elm Street, where access for wade anglers is much easier. You'll sometimes spot raft floaters who put in near that location and fish downstream to the railroad trestle near the I-91 overpass. When the water levels can accommodate wade anglers, get into the water and fish downstream through the run. Trout will be spotted rising in the deep pool just opposite Fred Mold Park. However, when water levels allow, it's best to ford the river

The Passumpsic River main stem at the Special Regulations section in St. Johnsbury. The Moose River joins the Passumpsic on the left of the photo.

Final falls at the East Barnet Hydro Plant, located just a mile upstream from Passumpsic River's confluence with the Connecticut River at Nine Island.

just downstream of the Moose River confluence and make your way upstream to cast from that side of the river.

Conventional tackle anglers often head to the deep pool below the final dam at East Barnet Hydro Plant, a mile upstream from the river's confluence with the Connecticut. The pool is deep and not all that easy to fly fish. Some even venture downstream to near Nine Island, a difficult trek if you're not floating. Nine Island, also favored by anglers from New Hampshire, is known for some large brown trout. You can easily reach the pool area below the dam by taking Comerford Dam Road off US 5. Cross the bridge upstream from the dam facility, then continue around the bend to shoulder parking next to the road. If you're fishing the downriver area near Comerford Dam Road, run up to Joe's Brook tributary, a nice little stream that holds brook trout.

Golden Stonefly Nymph and Maple Syrup Nymph/Emerger *Tied by Al Pitt*

Al said the Maple Syrup Nymph/Emerger seems to have been developed in Maine and is made in several sizes with or without a bead and with either red or yellow tails. In the larger sizes, it may be taken as a stonefly nymph or maybe an emerging *Hexagenia* mayfly.

Joe's Brook

Joe's Brook (aka Joes Brook) enters the Passumpsic River about 1.6 miles upstream from the Barnet Hydro Plant. Access it along Joe's Brook Road, the Greenbanks Hollow Covered Bridge crossing (off Greenbanks Hollow Road), and headwater reaches off Oneida Road. VF&WD stocks the stream with both rainbows and brook trout. About 400 brook trout are stocked annually from Oneida Road downstream to the Greenbanks Hollow Covered Bridge. About 300 rainbows are stocked from Brook Hill Road downstream to the Passumpsic. Wild brook trout inhabit areas around Morses Mills.

Joe's Brook at the Greenbanks Hollow Covered Bridge crossing. The bridge, located on Greenbanks Hollow Road, marks the downstream boundary where VF&WD stock brook trout. The upstream boundary is at Oneida Road.

From the US 5/Joe's Brook Road intersection close to Comerford Dam Road, drive along Joe's Brook Road 1.4 miles to Brook Hill Road. The stream comes into view along Joe's Brook Road a short distance from the intersection but flows largely away from the road with little access. That all changes at a nice section of stream, beginning at the Brook Road crossing. Park along the road shoulder on Brook Road, next to the bridge, or along Joe's Brook Road. Well-worn footpaths lead directly from the bridge into a picturesque area. At the bridge, Joe's Brook crashes down a short flume to gentle stretches inhabited by stocked rainbows. A couple of nice viewing benches are located along this reach, part of the Passumpsic Valley Land Trust area. The area is popular with swimmers and other recreational users.

After checking out the Brook Road reach, you'll want to head upstream to other interesting sections. From the Joe's Brook/Brook Hill Road intersection, continue 5.5 miles on Joe's Brook Road to Greenbanks Hollow Road. Take a left on Greenbanks Hollow Road then drive 1.9 miles to the historic restored covered bridge crossing in the "forgotten village" of Greenbanks Hollow. Now a ghost village historic site and town park along Joe's Brook, the area was once a thriving community. On December 14, 1885, a fire destroyed a woolen mill, store, and residences. Soon after, the remaining residents vanished. There's a picnic table at the site, and trailheads are located near the stream.

You reach Joe's Brook's upper most areas by taking Brook Road at the covered bridge, following the stream a couple of miles to a bridge crossing at Peacham Road. You'll encounter nice stream structure as you make your way upstream to Peacham Road. Cross Peacham Road and continue straight onto Harvey's Hollow Road. In about 1 mile you'll see Keiser Pond, which exists as a small outlet that joins Joe's Brook. Harvey's Hollow Road joins Keiser Pond Road. Follow Keiser Pond Road to Oneida Road, then make a sharp right onto Oneida Road. In about 2 miles, you'll get to a bridge crossing. Fish downstream with small flies for brook trout that inhabit the stream. The source of Joe's Brook, Joe's Pond's outlet, is located a couple of miles upstream from the Oneida Road bridge crossing.

White River

I don't know how many times I've driven past the small town of Granville while on the way to the Mad River that's born just up the road. Along the crossroads in Granville, a lovely little stream flows in an opposite direction from Mad River's journey along VT 100 north toward the Winooski River, contributing water that ends up in Lake Champlain. For a long time, I had no idea of the name of the little stream that flows across Granville's crossroads, then along VT 100/VT 107/VT 14 south toward the Connecticut River.

Indeed, Vermont's approximately 60-mile-long White River in its headwaters at Granville is a completely different stream in both appearance and other features when compared with the river's expansive lower reaches many miles downstream along its journey to its confluence with the Connecticut River at White River Junction. Among other tributaries, the White River picks up its three branches (Third Branch, Second Branch, and First Branch) from Bethel and downstream.

HEADWATER REACHES DOWNSTREAM TO ROCHESTER

White River is born in the Breadloaf Wilderness of the Green Mountain National Forest. It trickles out of Skylight Pond, just over the ridgeline slope that sends the Middle Branch of Middlebury River to the west to Otter Creek, then Lake Champlain. In the mountains, the White River quickly picks up Patterson Brook, a small brook trout tributary, before it

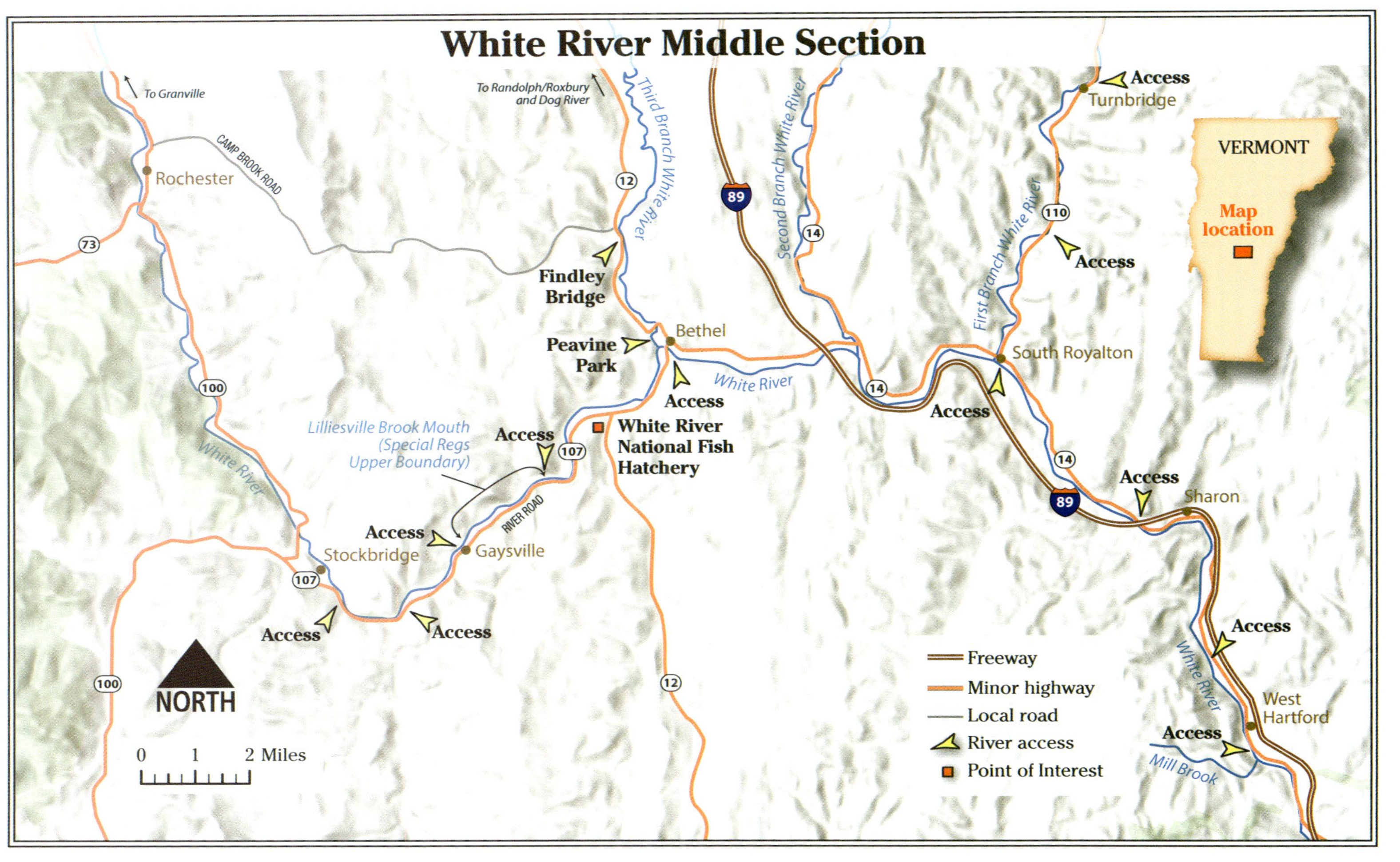
White River Middle Section
To Granville
To Randolph/Roxbury and Dog River
Rochester
CAMP BROOK ROAD
Third Branch White River
Second Branch White River
First Branch White River
Turnbridge
Access
VERMONT
Map location
Findley Bridge
Bethel
Peavine Park
Access
White River
South Royalton
Access
Access
Lilliesville Brook Mouth (Special Regs Upper Boundary)
Access
White River National Fish Hatchery
White River
RIVER ROAD
Access
Stockbridge
Gaysville
Sharon
Access
Access
Access
Access
West Hartford
White River
Access
Mill Brook
NORTH
0 1 2 Miles
Freeway
Minor highway
Local road
River access
Point of Interest
73
100
12
107
14
110
89

The Green Mountain National Forest Riverbend Observation Site is located along a VT 100 bridge crossing, just downstream from the town of Hancock. A couple of small picnic tables are located at the small parking lot directly next to the river. VALERIE VALLA

hits Granville. White River's upper reach and Patterson Brook are brook trout waters for sure, splendid little streams and a good place to cast a 3-weight rod with small bead-head flies into the little runs and pools.

White River's headwaters are easy to access by taking West Hill Road from the Granville crossroad area 1.2 miles to Patterson Brook's confluence. Bear left on FR 55 where the road splits in about 0.5 mile as it climbs the mountain. You'll notice that White River is largely inaccessible along a steep drop-off from the road, but once you get to Patterson Brook's confluence, access into the water is instant. Parking is not at issue along the road shoulder and area at the FR 50 (Patterson Brook Road)/FR55 intersection at Patterson Brook's confluence. Of course, an ambitious angler could park down at the West Hill Road/FR 55 intersection in Granville, easily get into the water, and then fish upstream into stretches that are largely inaccessible until you get to Patterson Brook.

High-elevation headwater brooks like the White River in Green Mountain National Forest are the perfect streams to head for when lower river reaches become too warm during periods of high ambient temperatures, putting trout into thermal-stress situations. Ethical anglers will leave the big river behind when temperatures rise to levels less than acceptable for angling, catch-and-release or not. During times of potential thermal stress, head upriver to the small water.

That's exactly what Pittsfield, Vermont, fly fisher Tim Major was doing when I ran into him a couple of years ago, parked alongside the White River's headwaters in Granville. Tim had recently returned from a successful trip on a nearby headwater brook trout stream, where he racked up nice numbers of fish. I know the stream well where Tim fished. We chatted about him being chided a bit about targeting smaller headwaters trout. Tim is one

Headwaters of the White River at the junction with Patterson Brook, upstream from Granville. The little stream tumbles through runs and pools that support native brook trout. Patterson Brook is also a brook trout stream. VALERIE VALLA

of those ethical Vermont fly anglers, like most really, who recognize when we should head to colder waters. He had more to say after we met that nice yet warm day in mid-July:

"Many people think of Vermont as a cold air trout haven, but the reality is that it can get just as hot from June to August as anywhere else. In such conditions, targeting trout in low basins can be deadly for them, even with the best intentions of catch-and-release. Fly anglers are often adamant about catch-and-release practices, but fishing for trout during hot days when ethical release isn't possible can do more harm than good. Instead, anglers should head to higher tributaries, where the forest canopies provide cooler air and the mountainside bedrock creates cooler flows, both creating a more sustainable environment for the fish. It's crucial to adapt our fishing practices to ensure the well-being of the trout. Vermont has experienced serious flooding in the last decade. The trout are making comebacks, but they deserve any bone we can throw to them to help them strive. As my father always said, 'If you hurt a trout, you better eat it.' His words play a huge part even when I'm deciding where to park my truck on a summer morning. There is absolutely nothing wrong with targeting dinner, but if your intentions are for ethical release fishing, your practices begin with the parking lot."

After the White River departs Granville along VT 100, it flows through a splendid, picturesque valley dotted with small Vermont towns and settlements. Access is available here and there at bridge crossings where you can get into the water. You'll note that at times the now-growing river is still very much like a small creek as it sometimes flows well away from the road and then closer to the road along other sections. Always respect posted property, but you'll find plenty of areas where you can get into your waders.

The Green Mountain National Forest Hancock Overlook Observation Site, 3.6 miles from Granville off VT 100, 0.5 mile after crossing the Granville-Hancock town line, is a small roadside pull-out parking area directly next to the stream. Small signage down on the field to the right of the parking area alerts visitors to stay on the public part of the lot. This reach is not what I'd call prime fly-fishing water, but willing trout can be had by fishing upstream from the lot.

A much better fly-fishing section is located downstream at the Green Mountain National Forest Riverbend Observation Site. It's located off VT 100 downstream from the settlement of Hancock (3.6 miles downstream from the Hancock Observation Site and 7 miles downstream from the Granville crossroads). Signage is at a VT 100 bridge crossing along the now much larger stream than what was encountered upstream in Granville and Hancock.

The Lions Club picnic area, located roadside just a mile down VT 100 from the Riverbend Observation Site, also provides easy access to the stream. An adventurist angler could park at the Riverbend site, fish a long stretch of the river that flows well removed from the road, then exit the stream at the Lions Club picnic area. After exiting the stream at the picnic area, walk the mile back to the Riverbend Observation Site.

ROCHESTER TO STOCKBRIDGE VT 100 BRIDGE CROSSING

This section isn't one of my favorites, but it's worth sampling at a few access points along road shoulder turnouts. After departing the quaint village of Rochester, keep your eyes open for road shoulder parking areas where you can access the river. The river soon vanishes from sight as it flows off in the distance along farm fields, but it reappears a few miles down the road. Keep your eyes open again for road shoulder parking. VT 100 soon crosses the river in Stockbridge. You'll notice the "National Forest Peavine Parking Access" area sign after crossing the bridge.

STOCKBRIDGE TO BETHEL—SPECIAL REGULATIONS SECTION

The reach between Stockbridge and Bethel is one of the more popular stretches for fly fishing. VF&WD typically stocks nearly 3,000 rainbow trout between the VT 100 bridge crossing in Stockbridge downstream to the Gaysville Bridge crossing. After crossing the VT 100 bridge over the White River, you'll soon skirt the Tweed River tributary as you approach VT 107, which you'll continue on for several miles to Bethel. Just after merging onto VT 107, you'll cross the Tweed River but soon again intercept the White River and get a good view of the river's character. About 2.7 miles beyond the VT 107 Tweed River bridge crossing, take a left onto Blackmer Boulevard. (From VT 107, you would have spotted the Blackmer Boulevard bridge crossing just before turning onto that road.) Park along the shoulder near the bridge and have a look at the nice river structure that's great for rainbows. The pool under the bridge is popular with swimmers during summer.

Continue on VT 107 less than 2 miles to another bridge crossing on Bridge Street in Gaysville, next to the Gaysville Post Office, where you can have another look at the river's continued nice water. The structured run upstream from the bridge takes a sharp turn before it slams into a rock ledge and deep pool. You're better off getting into the river on the downstream side of the bridge, made easy by the town of Stockbridge White River Access path down to the river. For the time being, you'll depart VT 107 for a quick 1.5-mile drive on River Road, your first right turn after crossing the bridge, to the mouth of Lilliesville Brook.

The mouth of Lilliesville Brook marks the upper extent of a Special Regulations section on the White Rive that's limited to artificial lures and flies only, and one harvested trout with 18-inch minimum length. Exact boundaries listed in the Vermont Fishing Guide & Regulations syllabus are "from the confluence with Lilliesville Brook in Stockbridge downstream 3.3 miles to 220 feet downstream of the confluence with Cleveland Brook in

Bridge crossing at Gaysville. Access to stretches downstream is made easy via the town of Stockbridge White River Access pathway, located at the bridge.

Bethel." You'll want to be casting Woolly Buggers and streamers off a 9-foot, 5-weight rod once you get into these lower river sections.

After scouting out the River Road stretch, return to VT 107 in Gaysville and continue toward Bethel. You'll pass the White River National Hatchery entrance road in about 3.5 miles from the VT 107/Bridge Street intersection, worth a stop if you have time. You'll cross Cleveland Brook, the lower boundary area of the Special Regulations section, a mile or so beyond that point. In another mile you'll cross the bridge into Bethel.

BETHEL TO SOUTH ROYALTON—THIRD BRANCH, SECOND BRANCH, AND FIRST BRANCH TRIBUTARIES

If you gaze upstream from the VT 107 bridge leading into Bethel, you'll spot the confluence of Third Branch of White River tributary. If you gaze downstream, you'll note nice water with runs and pools popular with anglers. Depending on the time of year, as well as water levels, you'll want to cast streamers and buggers into runs both upstream and downstream from Bethel Bridge. Access those stretches by taking the short gravel road that leads from VT 107 just before Bethel Bridge. Park at the end, at riverside.

The mouth of the Third Branch is a good place to get into brown trout. VF&WD stocks approximately 1,300 browns in the Third Branch, from the Randolph Sewage Treatment Plant several miles upstream to the confluence in Bethel. Immediately after crossing the bridge into Bethel, turn left onto Peavine Boulevard and follow the road to the bridge that leads across the Third Branch to Peavine Park along the confluence. The park has a gazebo and picnic tables—a nice place to sample the Third Branch as it pours into the White River.

Many upstream stretches along the Third Branch are tough to get direct access into, but check out the deep pool at Findley Bridge off VT 12. Get there by making a left turn after crossing the VT 107 bridge in Bethel. Then drive north 0.2 mile on Main Street to another bridge crossing that continues as VT 12. In 2.4 miles turn right onto Findley Bridge Road, which leads to the small truss bridge. Fish along the stretch that is located downstream from the bridge. If you continue north on VT 12 and get up to the East Braintree area (north of Randolph), try the headwaters of Ayres Brook, a small tributary. VF&WD has surveyed the stream in past years. Wild brook trout inhabit the upper reaches, and sometimes a few rainbows show up in the mix.

Getting back to the White River main stem, continue on VT 107 toward Royalton/South Royalton after crossing the bridge into Bethel. In just a couple of miles beyond Bethel, you'll pass under I-89, and in about 0.5 mile you'll cross the Second Branch of White River at the VT 107/VT 14 intersection.

You'll now be following VT 14, and in about 3.5 miles you'll cross the First Branch of White River. First Branch joins White River in South Royalton after flowing south along VT 110, passing through Chelsea, North Tunbridge, and South Tunbridge along the way. Access is a bit easier than on Second Branch. VF&WD stocks First Branch with about 800 rainbows and 800 brook trout annually between Chelsea Village and Turnbridge Fairgrounds. The fairgrounds are located a little over 5 miles from the VT 14/VT 110 intersection in Chelsea Village.

A nice side attraction found on First Branch, besides the rainbows and brook trout—and a few browns—is the series of a half dozen covered bridges that cross the stream. The Larkin Covered Bridge stretch, 8.3 miles north of the VT 14/VT 110 intersection, is nice. Try for brook trout in Chelsea by parking at the VT 110/Creamery Road intersection bridge crossing, about 13 miles from the VT 14/VT 100 intersection. You can also get into the water a short distance downstream from Chelsea, at Heath Recreation Park off VT 110.

SOUTH ROYALTON TO WHITE RIVER JUNCTION CONFLUENCE

Getting back to the main river at the VT 14/VT 110 intersection in South Royalton, you'll note the wide nature of the White River when you gaze upstream and downstream from the Chelsea Street bridge crossing. First Branch comes into the river just upstream from the bridge. A popular swimming area and structured water are located downstream from the bridge. A formal parking area used by recreational enthusiasts and anglers is located off the south side of the bridge.

As the White continues its path to the Connecticut River, it flows through Sharon, West Hartford, and Hartford before it hits White River Junction. While the river intercepts areas increasingly congested along downstream reaches, rainbows and bass can keep fly casters busy. VF&WD stocks between 4,000 and 6,000 rainbows between Bethel bridge down through the towns of Sharon and Hartford.

The river continues flowing through flat stretches but hits structured, rocky water along its path, perfect for rainbows. As you continue driving down VT 14, keep your eye out for informal road shoulder areas that can get you into the river. In addition to many informal road shoulder parking areas, as you progress down VT 14 you'll encounter a couple formal pull-offs just a couple of miles downstream from the VT 14/VT 110 intersection. In about 3 miles past the intersection, you'll first see the Town of Sharon River Access parking lot, marked by a small sign. The Ledges formal parking area, marked by a kiosk and signage, is a mile or so beyond the Town of Sharon River Access.

An interesting stretch can be accessed on the opposite side of the river from The Ledges by crossing the West Hartford Bridge (8 miles or so downriver from The Ledges). Cross the bridge and then turn right onto Pomfret Road. Drive about a mile to White River Lane, where you can park next to Mill Brook tributary. Note the large signage that describes the Mill Brook Culvert Fish Ladder Site, a 1995 project undertaken by Trout Unlimited that allows trout to reach spawning areas upstream. Walk upstream from the parking area along the dirt road to a rocky fast-water stretch. Plenty of other access is available along the White River, all the way to its mouth at White River Junction.

Black River

Born as an outlet from Black Pond in the town of Plymouth, the Black River continues 40 miles to the Connecticut River. The river drops more than 3,000 feet from Ludlow Mountain by the time it reaches Hoyts Landing in Springfield. The Black River supports both trout and warmwater species fishing opportunities. Over a dozen dams along the river and a number of ponds, lakes, and reservoirs impact the fishery. Black Pond (20 acres), Lake Amherst (83 acres), Echo Lake (100 acres), Rescue Lake (224 acres), and North Springfield Reservoir (96 acres) interrupt the Black River's flow. The impoundments are also popular with anglers.

Vermont's 2023 Department of Environmental Conservation *Basin 10 Tactical Basin Plan* provides good information relative to the Black River's fishery, both the river and its area ponds. Echo Lake, Lake Amherst, and Lake Rescue (also Lake Pauline, which it connects via a small dam) are a group of lakes commonly referred to as the Plymouth Lakes.

Valerie Valla taking a lunch break along the Special Regulations section at the Downer Covered Bridge. The Special Regulations stretch is blessed with fast-moving rainbow trout water during early season. It's best to fish this section during mid- or late May, before water levels drop and temperatures climb.

Echo Lake, located off VT 100 at Tyson, is a popular coldwater and warmwater species fishery and home to Camp Plymouth State Park, where camping is available. Echo and Amherst Lakes are stocked with rainbow and lake trout. Lake Rescue gets brown, lake, and rainbow trout. Boat access is available on the lakes via VF&WD areas.

Studies have shown that while Black River reaches upstream from Plymouth Lakes have water temperatures suitable for trout year-round in some sections, the cumulative effect of lake surface area exposure results in water temperatures downstream that cannot maintain a year-round wild trout fishery. Tributaries upstream from Cavendish Dam are managed for wild brook and brown trout. Although the thermally compromised reaches downstream from Cavendish Dam cannot maintain a wild trout fishery, the 4-mile-long Special Regulations section between the Howard Hill Road bridge and Downers Covered Bridge is probably the most popular fly-fishing section.

The put-and-take Special Regulations section is stocked with brown and rainbow trout. The best time of year to target rainbow trout in the Black River along the Special Regulations section is late May into early June. From late June and through the summer, the Black River becomes low and often very thin, with elevated water temperatures not conducive to successful fishing. VF&WD fisheries biologist Lael Will provided insight after I reported a small fish kill. "It's a put-and-take fishery, so we don't expect survival through the summer. The Black generally tracks air temperature, so they don't survive when we get hot days. More anglers are practicing catch-and-release, so you will likely see dead fish if they are

Dave and Pamela Bornholdt on the Special Regulations stretch in mid-June. Compared with water levels encountered during early spring, levels drop significantly in most years by the time late June arrives.

While most of the focus on the Black River is along the fast-moving stocked trout water that flows 4 miles through the Special Regulations section, slow-moving pools in other areas are available. Black River regular Stan Steen searches for rising trout at the Power Plant Road reach, downstream from Cavendish Dam and Cavendish Gorge.

not harvested. We do get some browns surviving during cool wet summers like last year, but in general we expect mortality."

The entire stretch can be accessed along VT 131 as it follows the Black River in a section downstream from Cavendish. Action can be had in the Black River system outside of the Special Regulations area.

While not in the Special Regulations mileage, an interesting fly-fishing section can be accessed off Power Plant Road, off VT 131 (Main Street) in Cavendish. As you make your way down Power Plant Road, you'll immediately notice on your left the remnants of a catastrophic rain event that occurred in July 2023. A small drainage that feeds the river blasted through the area, taking with it rusted culverts, debris, and even remains of an old vehicle—scattering it all along its path.

Park where Power Plant Road splits to the right. That's where I met Stan Steen as he was gearing up to fish one of his favorite pools downstream from Cavendish Dam. Stan and I marched our way through the debris field toward to the river. "It never appeared like this before the flood," Stan commented as we carefully dodged rusted metal. When we reached the river, Stan pointed out where trout like to hang out. "You'll run into a variety of bugs that emerge," he said. "March Browns were on the water recently, but other mayflies and caddis come off here too." Stan shared that he once caught a very large brown trout along the stretch.

Bornholdt Golden Squirrel *Tied by Mike Valla*

This multispecies fly entices both warmwater and coldwater species, rainbow trout included. It's tied using either a Keel hook, as shown here, or a regular streamer hook. Gray squirrel tail hair creates the wing.

Silver Opossum *Tied by Mike Valla*

The Silver Opossum's wing provides action as it drifts through Black River riffles and runs. Rainbows are enticed by the fly.

BLACK RIVER TRIBUTARIES

Several tributaries that enter the Black River deserve some mention, since they have received significant attention and study by VF&WD biologists. Vermont's 2023 Department of Environmental Conservation *Basin 10 Tactical Basin Plan* includes fisheries biologist Courtney Buckley's thorough description of tributaries; many sustain wild trout populations or serve as important spawning and nursery waters. Among the most important are Jewell Brook, Sanders Brook, Grant Brook in Ludlow, Twenty Mile Stream in Cavendish, Hubbard Brook in Windsor, and North Branch of Black River upstream from Ascutney Basin Road. Jewell Brook, a small stream that flows along VT 100 south of and then into the Black River in Ludlow, was described as one of the southern Vermont streams, along with Batten Kill, having "the healthiest, most robust trout population seen in southern Vermont." Both brown and brook trout inhabit the stream. For extensive information on these tributaries, readers should examine fisheries biologist Courtney Buckley's findings in the *Basin 10 Tactical Basin Plan*, available online.

Jewell Brook

A tributary of the Black River, Jewell Brook is a small stream born just a few miles south of Ludlow near Okemo State Forest. Anglers fishing the upper reaches of the West River and its tributaries, Jenny Coolidge Brook and Greendale Brook north of Weston, can quickly get over to Jewell Brook. Along its path, the stream's flow is interrupted at a tiny reservoir and then flows downstream a few miles, roughly paralleling VT 100, to its confluence with the Black River in Ludlow.

Jewell Brook is more interesting for its abundance of small trout than its meager total mileage. While not a major fly-fishing destination, it deserves a few casts before heading downstream to the Black River. VF&WD regularly surveys Jewell Creek. While the majority of brown trout are young-of-the-year, catchable sizes also inhabit the little stream. One of the challenges anglers have when desiring to cast a line along its small riffles and pools is the lack of easy access.

VF&WD regularly surveys the reach off Clark Road, a small bridge crossing located a short distance upstream from Ludlow. The little brook-like water upstream from the bridge holds colorful yet small wild brown trout. Parking presents a challenge. It's nearly

Jewell Brook at Dorsey Park reach in Ludlow VALERIE VALLA

impossible to find suitable parking along VT 100 road shoulders that are next to the stream. Parking can be very tight on the far side of the bridge directly next to the stream; make sure to avoid infringing on private property and dwellings in the area. Easier access is available at Dorsey Park, located in Ludlow less than a mile downstream from Clark Drive. Parking is available directly next to the stream, next to the recreation fields. Jewell Brook flows past dwellings along the side opposite the park, a downside of fishing the stream in Ludlow. Jewell Brook's confluence with the Black River is located behind the Ludlow Shopping Plaza, off Main Street (VT 103) in Ludlow. There's a somewhat steep yet short bank at the mouth, but it's manageable.

Another access point away from congested areas in Ludlow is located upstream at the Brooks Road/VT 100 crossing, a mile or so south of Clark Drive off VT 100. The stream is a very small brook in this area and requires using short rods. But you'll be fishing outside of the more congested downstream areas.

Bailey Brook

Like Jewell Brook, Bailey Brook isn't a major fly-fishing destination, but it's worth a few casts if fishing the Black River near Cavendish to its south or after fishing the Knapp Ponds. The best mileage is located off Baileys Mills Road, which intersects VT 106 at the small crossroads of Hammondsville in the Reading area. Baileys Mills Road is about 7 miles north of VT 106/VT 131 (just east of Cavendish). Bailey Brook combines with Mill Brook in the Hammondsville area, then sends its combined water southeasterly to Windsor, where it dumps its water into Mill Pond and then the Connecticut River.

VF&WD regularly monitors Bailey Brook's water temperatures and surveys its trout population. In the 2023 *Basin 10 Tactical Basin Plan*, VF&WD fisheries biologist Courtney

Bailey Brook is a small stream located in the Reading area, northeast of Ludlow and Cavendish. It's known for its small trout. VALERIE VALLA

Buckley reported long-term brook trout population data. Fisheries biologist Lael Will's 2022 *Inland Salmonid Report* (also available online) also presented interesting population data. Plenty of young-of-the-year brook trout and brown trout inhabit the stream, combined with a lower number of catchable-size fish, as is always the case.

The best section to fish is about 0.6 mile west of VT 106/Baileys Mills Road. This reach is located where VF&WD monitors water temperatures and surveys trout populations. It's nicely shaded. You'll find a small parking area at an old bridge crossing next to a fire hydrant next to the stream. Make sure you're not blocking the hydrant area. An alternative is to park along Baileys Mills Road.

The stretch runs about 0.5 mile upstream from the parking area before the brook flows out of a more open area with dwellings situated near the water. Valerie and I began sampling the brook at the old crossing. By late August, the stream runs cold and very clear, perfect for hip boots or wet-wading. The first pool looked very promising, and I half expected to spot a trout. Valerie spotted a very nice brook trout, maybe 10 inches, that began to dart around as I moved upstream beyond her vantage point. A short rod is perfect for this reach.

A few stretches worth exploring can be found downstream, where water is larger from inflow from Mills Brook, which joins Bailey Brook before Bailey Mills Road intersects VT 106. Heading south on VT 106 about 0.8 mile from the VT 106/Baileys Mills Road intersection, you'll encounter road shoulder parking that provides access to the stream, now Mill Brook. If fishing the area in early spring and you're able to haul a kayak or canoe, consider running up to the nearby Knapp Ponds, just up the road from Cavendish.

Knapp Pond #1 and Knapp Pond #2

The Knapp Ponds, nestled in the 1,249-acre Knapp Brook Wildlife Management Area, are man-made impoundments operated by VF&WD. They're located just up the road

Knapp Pond #2, the connecting pond to Knapp Pond #1, is a stocked brook trout pond in the vicinity of the Black River near Cavendish. In a few ways, Knapp Pond #2 is somewhat nicer than Knapp Pond #1.

from Cavendish, making the two ponds an excellent addition for fly anglers roaming the Black River during spring. Locate to Tarbell Hill Road (almost directly across VT 131 from the Howard Hill Road crossing on Black River, the upper boundary of the Special Regulations section).

From the Tarbell Hill Road/VT 131 intersection in Cavendish, head for 1426 Knapp Pond Road. The first of the two ponds you'll encounter, marked by formal VF&WD signage, is Knapp Pond #1. The large parking lot and kiosk are located directly next to the boat launch, making for easy access into the 25-acre pond, created by an earthen dam completed in 1958. Like its sister connecting pond, no structures surround the pond and swimming is not allowed, making for very pleasant fishing. If you're interested in targeting rainbow trout, VF&WD stocks about 6,000 rainbows annually, ranging in size from 9 to 17 inches, during April. Brook trout are not stocked in Knapp Pond #1.

If you're interested in targeting brook trout, head up the road a short distance to another large parking area and boat launch at Knapp Pond #2. Nearly 2,500 brook trout, ranging in size from 9 to 14 inches, are stocked annually in April. Completed in 1961, the 35-acre pond provides pleasant fishing. I most enjoy larger Knapp #2 Pond for its configuration. From the vantage point of the launch, the pond at midpoint elbows to the right, out of sight. Off in the distance, two islands, one quite small, add to the aesthetics. Cast or troll streamers along the larger of the islands, the first that comes into view as you paddle away from the launch.

You won't encounter the variety of fish species you might expect to latch onto your flies in the ponds. During summer, with its lily pads and cover, you might expect panfish or a largemouth bass to hammer small popper flies. In the 2023 *Basin 10 Tactical Basin Plan*, VF&WD fisheries biologist Courtney Buckley reported interesting findings concerning the Knapp Ponds.

Knapp Pond #1, the smaller of the two connecting ponds in the Cavendish area, is easily accessed via a formal boat launch at a large parking area.

"The Knapp Ponds have low species diversity including Brown Bullhead and Fathead Minnow. Fathead Minnows were discovered in both Knapp Ponds in 2007, likely following an illegal introduction, and the use of baitfish for fishing, either alive or dead, was prohibited at the Knapp Ponds until 2011. Knapp Ponds were closed seasonally until 2022 when VF&WD opened a number of lakes and ponds to ice fishing. Anglers were able to catch a number of brook trout in both ponds, suggesting that brook trout are surviving the summer and fall, which was not known previously. More investigation into potential holdover of stocked fish is needed to determine whether management changes should be made to accommodate increased angling pressure in the winter."

Wells River

Wells River begins its 15-mile journey in Groton, Vermont, at Ricker Pond, picking up Red Brook, South Branch, and North Branch, joining the Connecticut River at its namesake village of Wells River. Wild brook trout inhabit the upper reaches near Groton, while rainbow trout can be found in lower sections. VF&WD stock a modest number of rainbows—around 600—from South Ryegate (downstream from Groton) to its mouth. I'm told by staff at Tuttle's Family Diner in Wells River (a great place to stop for lunch) that stockings occur during mid-May at the bridge crossing right in town.

While access is easy at a formal VF&WD parking area, marked by signage off US 302 (1.3 miles upstream from the US 302/US 5 intersection in the village of Wells River), I don't consider it prime fly-fishing water, but it's worth a look. The river at this access, called "The Hole in the Wall" by locals, tumbles through an area that was once the site of a brick factory. Remnants of a dilapidated dam are strewn in the water, although pools favored

Wells River, just upstream from a VF&WD access point off US 302. Access can be difficult from this point upstream into better reaches that flow through its headwaters.

The Wells River begins at Ricker Pond, just a couple of miles upstream from West Groton. The shallow pond is inhabited by smallmouth and largemouth bass. Small watercraft can be rented at Ricker Pond State Park.

more by swimmers than trout are in sight. The sandy-bottom river just upstream from the access area is typically shallow and unappealing, although one brief run enters the long flat.

A couple of stretches in its headwaters, where Wells River tumbles out of Ricker Pond (about 1.6 miles up VT 232 from the VT 232/US 302 intersection in West Groton), are worth sampling. One is located in an open area just downstream from the entrance road into Ricker Pond. Consider a warmwater species experience at Ricker Pond if you're up that way. Ricker State Park has more than 20 RV/tent sites and a couple of cabins. The shallow 92-acre pond holds smallmouth and largemouth bass. A boat launch is available at the park, where kayaks and canoes can be rented.

SOUTH BRANCH WELLS RIVER

Given a choice between fishing the Wells River from Ricker Pond into Wells River village and fishing the South Branch of Wells River that flows close by, I'd take the latter. While the stream is small, especially in its headwaters, it is inhabited by wild brook trout—from french fry–size to an occasional dollar bill–size fish.

The South Branch of Wells River flows from the Noyes Pond outlet then meanders downstream as a rather small brook, eventually coming into view along Seyon Pond Road. The wonderful little brook flows through areas where dwellings are situated across the road from the stream, but access is available here and there in sections that are not posted. Always respect private property. Once it gets down to the US 302 area, access is not much of a problem at bridge crossings, where the stream tumbles through faster water. Park near Westville Road where it joins US 302 and fish the pocketwater with small nymphs. From the bridge crossing at US 302, the stream flows about 1.5 miles to its junction with the main stem. You'll be doing some small boulder and rock hopping as you fish down along US 302.

The South Branch of Wells River exits Noyes Pond in Groton as a small brook and flows toward Seyon Pond Road then along US 302 to its confluence with the growing main stem.

Noyes Pond (aka Seyon Pond)

Although officially named Noyes Pond on USGS topographical maps, most regulars call it Seyon Pond. The pond is located at Seyon Ranch, a state park (only someone like my wife, Valerie, could instantly recognize "Seyon" as "Noyes" spelled backward). Since that time, the pond has been highly protected as one of the state's prized brook trout ponds. Barbless hooks (or barbs that have been smashed down) are required,. A day-use fee is required to fish the pond. Personal floating craft are not allowed, and anglers are prohibited from fishing from shore. A small number of rowboats are available for rent by the hour, half day, or full day.

A variety of fly patterns are effective for taking Noyes Pond brook trout. Regulars have their own favorite patterns. Valerie and I ran into Burlington-area fly angler Ed Collins and his lovely wife, Laurie, on the pond a while back. In between swatting the black-flies that are out looking for a blood meal during mid-May, we chatted about favorite fly patterns. Ed pulled a handsome spinner fly from his fly box and dropped it in my hand. "This is my favorite when the *Hexagenia* hatch is on the water," he said. "The S&R Hex Spinner was created in Michigan by Tim Neal." Ed also likes an Adams Parachute when fishing the pond.

My first experience fishing the *Hexagenia* on Noyes was during a pleasant late-June evening fishing with Noyes Pond regular Leighton Wass, author of the superb *Fly Fishing the Hex Hatch* (2022), which he gifted to me weeks before. Leighton is a retired school teacher who taught in Plainfield, a small community on the far side of Spruce Mountain; its peak practically casts a shadow on Noyes Pond. Now in his early 80s, Leighton wrote:

Noyes Pond and the historic Seyon Lodge at Seyon Ranch, nestled in the 27,000-acre Groton State Forest, is located at 2967 Seyon Pond Road in Groton. Once a private hunting and fishing facility, the property was acquired by the State of Vermont in 1969. Noyes is the sole fly fishing–only pond in Vermont.

An information kiosk that describes the history of Seyon Ranch and its pond is located at the parking area next to Seyon Lodge.

"I not only want anglers to know *how* to fish the Hex hatch but also want them to know what makes this mayfly tick."

Leighton's knowledge of the bug, particularly on Vermont's ponds, is vast. Despite my own *Hexagenia* experiences reaching back to the early 1970s with *Hexagenia atrocaudata* on the Chenango River in New York and Hex hatches on that state's Adirondack Mountain remote brook trout ponds in later years, I still learned a heck of a lot from Leighton's book. Correspondence over time also provided insight into Vermont's best *Hexagenia* ponds and lakes.

Hexagenia limbata, one of Noyes Pond's most anticipated hatches, emerges sometime in late June.

Leighton came strolling down to the pond that evening, fully equipped with three rigged fly rods, a long-handled landing net, and other necessities. "These patterns work well," he explained, pointing to a few he had ready. His book also shows fly patterns used by others. One thing quite apparent is the vast array of styles and features that will take fish.

Valerie and I followed Leighton to our boats, ready and reserved for our use. (It's highly recommended that you contact Seyon Lodge by phone prior to your planned arrival, as the few craft available for rent "go fast during Hex time," Cheryl, the lodge's manager, told me.)

One bit of advice I didn't expect from Leighton while fishing Noyes Pond came before I threw my first cast. "Watch the loons! When you get a fish on, pay quick attention—they'll follow your fish as you're reeling it in then snatch it as a meal. They'll even swallow sizable trout!" Leighton rowed his boat up the left shoreline (where most others head too).

Valerie and I followed with our boat, keeping our distance from Leighton's position.

Just so my old eyes could see the darn fly on the water, I fished a clipped white deer hair–bodied Coffin Fly, a Catskill pattern that goes back decades. The brook trout loved it; maybe the fish could also spot the fly easier. I told Leighton about my success with the Coffin Fly and Kennebago Muddlers. My guess is, as I also mentioned to Leighton, open any angler's fly box during the Hex hatch and you'll likely be impressed by the variety of flies that are effective. I've used a dry fly pattern called "Big Fish Fly" during *Hexagenia* hatches on Adirondack Mountain ponds and lakes in New York state, largely because I can see the thing floating during last-light. Leighton chimed in:

Leighton Wass on Noyes Pond. In his *Fly Fishing the Hex Hatch*, Leighton shares his knowledge of the pond, its fish, its *Hexagenia* hatch, and the history of Seyon Ranch. LEIGHTON WASS

"I would imagine that you are right about the Coffin Fly. I have used the Kennebago Muddler but not often, and only as an emerger-style on the surface. Just recently, I tied up The Usual, a fly you are well aware of! About a week ago, while at Seyon, things became absolutely still around 7 p.m. No fish rising at all and calm. So I decided to put one on, kind of a 'stab in the dark' type decision. Second cast and blam! The biggest trout of the night, 12 inches, hit it like a freight train. I caught two more right off with it until it wouldn't float anymore. I should have ginked it up *before* using."

After netting a few 8-inch fish, a heavier brook trout took the fly. My 8½-foot, 5-weight arched. Val grabbed the camera, but within a few seconds it was gone. Then a big loon popped up out of the water, with the fish in its mouth—along with the only Coffin Fly I had in my box. Leighton and I had both rigged one of our rods with a sink-tip line. It's a good idea to try subsurface emerger patterns either during the hatch or before the hatch gets underway.

One subsurface pattern that brought me strikes that evening was a Kennebago Muddler, the fly Leighton said he also fished in the past. Fish hit both the natural color and a yellow version I first obtained years ago from Brett Damm, up in Rangeley, Maine. It sinks well because the spun deer hair head is clipped very short. Leighton's success the evening we fished together came from casting a rubber-legged *Hexagenia* pattern.

Hexagenia annual hatches aside, I asked Leighton about another mayfly emergence on Noyes Pond that another angler fishing Noyes that evening mentioned to me—the Green Drake hatch. "I'm guessing he was referring to the DARK Green Drake hatch that occurs

S & R Hexagenia *Tied by Ed Collins, created by Tim Neal*

Kennabago Muddlers *Tied by Brett Damm*

Big Fish Fly *Tied by Ralph Graves*

Pheasant Wing Hexagenia *Tied by Mike Valla*

at the end of May," Leighton said. "I refer to it as the Recurve Hatch in my book. It is *Litobrancha recurvata* and was once classified in the *Hexagenia* genus. It is a great hatch to fish, since it occurs during the afternoon and, like its cousin *Litobrancha*, is HUGE."

Martins Pond

While Noyes Pond doesn't allow private watercraft on the water, Martins Pond, another brook trout pond in the area, provides a launch area for those wishing to fish from their own boats. Martins, like Noyes, is also a Special Regulations trout pond. Martins Pond has been studied by VF&WD to ascertain best management for the brook trout population. Wild brook trout swim in the pond, but the numbers are not high enough to sustain a robust angling experience. Unlike Noyes Pond, a completely wild brook trout fishery, Martins Pond receives supplemental trout stockings. Biologists have found that wild brook trout numbers have been declining consistently since 2013.

In previous years, Martins was stocked with a few hundred brook trout ranging from 8 to almost 15 inches. However, in 2024 Martins Pond was stocked with 800 rainbow trout up to 15 inches. VF&WD biologist Jud Kratzer kindly provided the 2022–23 *Inland Fisheries* report he prepared. The report raised the option of stocking rainbows instead of triploid brook trout. It had to do with the discovery that although triploid female brook trout are sterile, the sterile triploid males sometimes exhibit spawning coloration and behavior that attract the attention of wild female brook trout. In the report, Jud wrote:

"It is possible that the stocking of male triploid Brook Trout is contributing to declines in wild Brook Trout abundance in Martins Pond. These stocked males are larger than many of the wild males, and so they may be able to outcompete wild males for mates. Females

Martins Pond, located not far from Noyes Pond, is another option for those seeking a small-pond brook trout fishing experience. It's located in the Peacham area, northeast of Noyes Pond.

that would choose to mate with these stocked males would be wasting their eggs, because these males are unable to fertilize them. If the number and size of stocked males is large relative to the number of wild males, this waste of eggs by wild females could have devastating consequences for the wild population. In contrast to the decline observed at Martins Pond, the wild Brook Trout catch rate has been relatively consistent at Jobs Pond—it has not been stocked since 2011." Jud pointed out that Martins wasn't always a brook trout pond to begin with. Prior to the 1958 reclamation, chain pickerel and perch inhabited the pond.

Getting to Martins Pond can sometimes get confusing. The first time I visited Martins, I had a heck of a time finding it. Head to Peacham and get to the Peacham Store, located at 643 Bayley-Hazen Road, a good starting point (and a nice place to get breakfast). Martins Pond is located about 3 miles from that point. At the corner of Church Street and Bayley-Hazen Road, take Church Street to Green Bay Loop Road. Once you're on Green Bay Loop, you'll drive a distance and wonder if you took a wrong turn. Continue until you see formal VF&WD signage that leads to the boat launch access.

I use my Hornbeck craft when fishing Martins Pond. You'll encounter other recreational boaters using everything from pontoon boats to kayaks on the pond. A number of cottages and small dwellings are located on the pond, mostly on the north shore. It's a much different experience compared with Jobs Pond, Noyes Pond, or other small brook trout ponds.

Waits River

Another stream born in the town of Groton is the Waits River. The most popular fly-fishing stretch on the Waits River is the stocked section between East Corinth and the village of Bradford. While the upper part of the watershed tributaries supports natural trout habitat, the lower main stem relies heavily on supplemental rainbow trout stockings. The 2020 *Basin 14 Tactical Basin Plan* stated that despite many years of stocking, the Waits River doesn't support naturalized rainbow trout populations. VF&WD typically stocks about 2,500 rainbows. Begin your outing at the Village Road/Waits River Road (VT 25) intersection at the East Corinth General Store. The folks who operate the store—a good place to pick up food and drinks—are very friendly and helpful. Stockings generally begin directly behind the general store (at the confluence with Tabor Branch tributary, a brook trout stream) and continue downstream along VT 25 to Bradford Dam.

The first good access is located along VT 25 0.4 mile from the general store. Other access areas are found downstream at bridge crossings and road shoulder turnouts. As you're continuing your way downstream, you'll note that some sections are wide open, shallow, and with little structure. However, keep your eyes open for areas that have better runs and pools. One nice stretch is located about 6 miles downstream from the general store, not far upstream from Bradford. You can park along the road shoulder. Freshly stocked rainbows are not too picky about what flies they'll eat. I like small streamers with a small amount of flash, but Woolly Buggers can do the trick too.

Waits River, a 25-mile-long river that joins the Connecticut River at the Village of Bradford, begins its life in Caledonia County in the town of Groton, about 20 miles south of Wells River. A good strategy if you're in the general area is to try fishing both streams. VALERIE VALLA

Ompompanoosuc River

Located about 10 miles north of White River Junction and about 15 miles south of the Waits River at Bradford, the Ompompanoosuc River is as interesting as its name is a mouthful. The river originates from two branches, the East and the West. Both branches flow southeasterly and converge just upstream from Union Village Dam. After exiting Union Village Dam, the fishery changes along its final nearly 4 miles before it enters the Connecticut River.

For trout fishing, both stocked rainbows and natural brook trout (inhabiting the branches' headwaters), you'll want to first head to the Union Village area. From there you'll head into Thetford for East Branch experiences and north to its headwaters. For the West Branch reaches, you'll head 8 to 10 miles west of Thetford to the South Strafford and Strafford area. You'll also want to check out the East Branch headwater for natural brook trout.

It's very easy to sample one of the branches then quickly locate to the other after a short drive along country roads that connect the streams. If driving up I-91, take exit 14 for Union Village, then head about a couple miles up VT 113 to Thetford. Your first stop will be the East Branch.

Ompompanoosuc River—East Branch

The Ompompanoosuc River's East Branch begins 14 miles or so upstream from Thetford, near Vershire. The small stream then flows along VT 113, past Mill Village, then West Fairlee. Another 6 miles or so downstream from West Fairlee, VT 113 intersects Tucker Hill Road in Thetford. Drive about a mile down Tucker Hill Road to the covered bridge crossing. This will be your first stop when fishing the East Branch, with its easy access

East Branch of the Ompompanoosuc River along the stocked section in Thetford, downstream from the covered bridge on Tucker Hill Road. The stocked section runs between Tucker Hill Road and Union Village Dam.

along the Union Village Dam Multiuse Trail Network made possible with cooperation of the US Army Corps of Engineers.

The covered bridge at Tucker Hill Road marks the upper boundary of a section along the East Branch that is stocked by VF&WD. The stocked rainbow trout stretch runs downstream from the covered bridge to Union Village Dam, the lower boundary. About 600 rainbow trout are stocked annually by the end of May along the 3 miles or so from the covered bridge to Union Village Dam. The access point might be confusing once you arrive at the covered bridge; you'll wonder just how you can get down into the water. A small parking area is located on the far side of the bridge, hidden off the main road. After crossing the bridge, you'll see a mailbox on the left at a dirt drive. Take the dirt drive just a short distance to the small parking area that provides access to the multiuse trail that follows the East Branch downstream. A well-worn path leads from the parking area to the stream. You'll encounter fast water tumbling at what was once a millsite.

Another area along the stocked trout section can be reached by taking Buzzell Bridge Road off VT 113, located less than 0.5 mile south of the VT 113/Tucker Hill Road intersection. A short drive down the road leads to a large parking area at streamside, where access into the river is easy. This section is also part of the Union Village Dam Multiuse Trail Network. The road at this point is sometimes gated off, but you can park at the lot and walk beyond that point to fish other areas of the reach.

The headwaters of the East Branch hold brook trout. The water upstream from West Fairlee, above Ricker Bridge Road, is small water but pleasant to fish during summer. Most any small dry fly lightly skittered over the small pools will bring the mostly fry-size brookies to your hand. However, dollar bill–size fish are there too. I use my 6- or 7-foot, 3-weight cane rods along the headwater reaches. Look for places along VT 113 where you can pull over and get into the water.

Natural brook trout inhabit the small water upstream from West Fairlee. Small pools and nice little runs hold fish. Most are small brook trout, with an occasional dollar bill–size fish in the mix.

Ompompanoosuc River—West Branch

Isolated stretches on the West Branch's headwaters, with a population of small yet wild brook trout, are much more appealing to me than either the marginal downstream reaches or the short, stocked section. However, larger reaches below Strafford are worth exploring. Local anglers report catching trout downriver. Head for the Strafford area, northwest of Thetford, to sample the West Branch of Ompompanoosuc River. Multiple routes will get you there, depending on your starting point.

An approach from Thetford via Tucker Hill Road will get you within quick range of the river. From the covered bridge on Tucker Hill Road, the location that delivered you to the East Branch, drive about 1.5 miles toward VT 132. Just before the Tucker Hill Road/VT 132 intersection, you'll see a road on the left, part of the Union Village Dam Multiuse Trail Network, that leads along the river.

Most of the time you can drive down the road to a bridge crossing. If the road is gated, just stroll down toward the crossing. Parking is easy along the Tucker Hill Road shoulder. You'll often encounter hikers parking along the road. Note the fabulous pools that come into sight pretty quickly. Keep in mind that the beautiful pools attract more swimmers than anglers during summer. The lower river reach doesn't have large numbers of trout, but local anglers report getting into them that far from Strafford.

Another access area, which takes a bit of effort to get into the water, is located just around the corner at a VT 132 bridge crossing. Continue down Tucker Hill Road a short distance to VT 132, then turn left to the bridge. Park on the opposite side of the bridge, at the Gove Hill Road intersection. If you're up to it, carefully trek down the hill to the water on the upstream side of the bridge (watch out for the poison ivy!). Fish the riffles and pocketwater.

Lower reach of the West Branch, along the Union Village Multiuse Trail Network. While brook trout can be encountered along this section, most of the fish are taken upstream along VT 132 near Strafford Village, where VF&WD stock fish annually.

Headwaters a few miles upstream from Strafford Village support small yet wild brook trout. Use short rods and small dry flies, twitched through pockets and small hiding places. VALERIE VALLA

Much easier sections to access are located up VT 132 in the South Strafford and Strafford area. Park along the road shoulder at points that are more manageable for accessing the stream. VF&WD stocks around 400 brook trout along the stretch between Strafford Village and the Strafford Fire Association building in South Strafford. Access the stream just up VT 132 from the Fire Association building, at the George I Varney baseball field. It's next to the recycling center. At the far rear-right of the baseball field, next to the stream, a set of hefty concrete stairs leads directly down to the water.

If you're after an enjoyable experience dabbling for small yet wild brook trout, head north of Stafford on Justin Morrell Memorial Highway about 2 miles to Taylor Valley Road. Drive another 0.7 mile to a small bridge crossing on Mundel Road, and park next to the bridge for easy access both upstream and downstream. It's a nice place to wet-wade during the heat of summer. You'll have small french fry–size brookies that time of year, eager to snip at size 16 dry flies. Elk Hair Caddis work well. A nice dollar bill–size brookie ate one of my favorite brook trout dry flies—a Lime Sally Stonefly.

Lime Sally *Tied by Mike Valla*

Ottauquechee River

As is true with other Vermont trout streams, the Ottauquechee River is a tale of two streams: stocked downstream reaches and wild trout upstream reaches and tributaries. Vermont's Department of Environmental Conservation 2023 *Basin 10 Tactical Basin Plan* described the river this way:

> Naturally reproducing (wild) native brook trout remain common in colder, higher elevation tributaries and within the main stem above West Bridgewater. Wild populations of naturalized brown trout and rainbow trout are less common but are present within individual tributaries and main stem reaches. Wild trout populations within the main stem below West Bridgewater are limited, likely due to high summer water temperatures and poor overall habitat quality.

Beginning in the town of Killington, the Ottauquechee River flows nearly 42 miles to its confluence with the Connecticut River, a few miles south of White River Junction. Along its traverse, the river flows through the towns of Bridgewater and Woodstock, then pushes past Woodstock to the village of Quechee and its magnificent Quechee Gorge, not far upstream from the confluence.

HEADWATERS IN THE TOWN OF KILLINGTON TO BRIDGEWATER CORNERS

Beginning in Killington near the US 4/VT 100 intersection, the Ottauquechee River headwaters pick up Kent Brook, a small stream that enters and exits Kent Pond, a public access pond stocked with both brook and rainbow trout. While some sections of Kent Brook were

The Bridgewater Park stretch provides easy access to the river. You'll likely be targeting stocked rainbow trout along its runs and pools. VALERIE VALLA

identified in 2021 as having potential for Class B-(1) recreational fishing, its outflow from Kent Pond isn't as impressive as its popular Thundering Falls. Located a short driving distance from the US 4/VT 100 intersection via either Thundering Brook Road or River Forks Road, the falls are said to be the sixth highest in Vermont. The 900-foot boardwalk that leads to the falls off River Forks Road crosses the brook. Viewing the brook from the boardwalk provides a pretty good indication of its less-than-impressive features and flow.

Roaring Brook tributary, located at the US 4 (VT 100)/River Forks Road intersection, is much more impressive; wild brookies inhabit the tumbling little stream. Driving south a little over 7 miles from the US 4/VT 100 intersection on US 4, you'll almost miss realizing that you passed over the stream.

Roaring Brook flows under a double parallel bridge crossing at that point, one under US 4 (VT 100) and the other a small concrete bridge off River Forks Road. Park along the lower bridge and have a look. Just for grins, while we were driving past the stream on our way home from a recent trip, I stood on the bridge and laid a long cast down into the water below. A dollar bill–size brook trout smacked the Elk Hair Caddis as I grinned to my wife Valerie: "See, I told you so!" In a way, I was glad the fish broke off so I didn't have to climb down into the stream to release it. A return trip resulted in some nice brookies.

Roaring Brook flows a relatively short distance downstream from the bridge crossing before it joins the Ottauquechee River. With a little effort, you can get into Roaring Brook upstream from the US 4 (VT 100) crossing. The upstream section here changed dramatically after the 2011 storm and the epic 2023 rain event. Lots of bank stabilization boulders have been placed. You'll want to be careful wading up the brook with its somewhat steep grade as it flows down from the mountain area.

The Ottauquechee River flows away from US 4 (VT 100) as it makes its way toward Bridgewater Corners but comes back into sight at US 4 (VT 100) bridge crossings and other side road crossings. In just over 4 miles south from the US 4 (VT 100)/River Forks Road intersection upstream, the river reaches the US 4 (VT 100) intersection at the general store and gas station. A nice run at the intersection is worth fishing. VT 100 then departs US 4 and heads in a different direction.

The Ottauquechee continues its traverse generally easterly then northeasterly, roughly following US 4, to an intersection with VT 100A a few miles down the road at Bridgewater Corners. The VT 100A/US 4 intersection in Bridgewater Corners is an important landmark because it represents the upper boundary of the long, more than 11-mile stretch that annually receives rainbow trout stockings. VF&WD typically stocks 2,500 rainbows between VT 100A/US 4 and the covered bridge at Taftsville (downstream from Woodstock).

Before exploring downstream reaches via US 4, cross the truss bridge in Bridgewater Corners and take a quick left onto River Road (a dirt road). River Road follows the opposite side of the river and provides easy access to a few runs and trout-holding water. This reach provides nice water to swing small streamers for stocked rainbows. River Road continues a short distance before it eventually dead-ends. After fishing River Road, return to US 4 and head farther downstream to other access points.

BRIDGEWATER CORNERS TO WOODSTOCK

The next access point is at Bridgewater Park. The park entrance is off US 4, 1.3 miles from the VT 100A/US 4 intersection. The river is located in the rear of the park and most easily accessed by walking to the left side of the fields then trekking a short distance down to the water. Pleasant water flows through an area where an 8-foot, 4-weight rod will serve you fine. Stocked rainbows will hit small bright streamers.

Lincoln Covered Bridge reach, not far upstream from Woodstock Village, is part of the rainbow trout–stocked mileage that runs from Bridgewater Corners to downstream from Woodstock. VALERIE VALLA

After trying Bridgewater Park, head down US 4 to the Lincoln Covered Bridge stretch in the town of Woodstock. It's located off US 4, just short of 4 miles downstream from Bridgewater Park. Drive across the bridge and park at a convenient road shoulder parking spot. A footpath leads directly down to the water. The stretch is often quite shallow during summer and warms up quite a bit by then. The best time to fish this stretch is May and early June.

After fishing Lincoln Covered Bridge, get back on US 4. You'll shortly pass through Woodstock Village, a bustling resort area known better for its restaurants, unique lodging, retail shops, and events that happen all year. Woodstock becomes pretty congested with visitors during the tourist height of late spring into fall. While you'll see some neat water at bridge crossings in the village itself, I've always stayed clear of those areas, given the congestion. Interesting warmwater fishing opportunities downstream from the gorge are no doubt there for exploration, but I've never fished down that far to find out what's available.

Colton Pond and Kent Pond

If you're lugging a small watercraft while exploring Ottauquechee River's headwaters and tributaries near Killington, stop by Kent or Colton Pond. Trout fishing is best during April until mid-May. Trout are stocked in both ponds in early spring. Both ponds also support a warmwater fishery. The nice thing about both ponds, besides the fisheries, is their proximity to each other. Colton Pond's formal VF&WD launch is less than 2 miles from Kent Pond's launch. Both VF&WD access points are off VT 100. Kent Pond has an additional access point on the east side of the pond.

Colton Pond is the smaller of two neighboring ponds located near Killington, Vermont. Access is easy and direct at the formal VF&WD boat launch off VT 100. VALERIE VALLA

COLTON POND

Colton Pond is the smaller and shallower of the two ponds. The 26-acre pond is only 11 feet deep at its maximum, more suited to bass fishing. However, in very early season between April and mid-May, VF&WD stocks about 1,000 brook trout. A couple hundred 2-year-old, 14-inch fish are in the mix. If you're targeting trout, very early season is the time you'll want to launch your small craft. Largemouth bass inhabit the pond, the species you'll want to target later in the season.

During summer, expect to share the water with recreational kayakers and others who might be visiting the Killington area or who are camping at the adjacent 285-acre Gifford Woods State Park. Gifford Woods is known for having one of the few remaining old-growth stands of hardwood trees in Vermont. Four cabins and 21 RV/tent sites are available. I prefer Colton over Kent because it's in an area that provides the feel of an isolated, remote pond. Kent Pond is much busier during summer, largely because of its size—one reason you'll expect company.

KENT POND

From the VT 100/US 4 intersection in Killington, drive about 0.4 mile north on VT 100 to the Kent Pond access road on the right, marked by a VF&WD sign. The parking area at the launch is large and located directly next to the water. In addition to stocked trout, 101-acre Kent Pond is inhabited by largemouth bass. Submerged rocks, brush, and obstructions provide good cover for bass.

If you're targeting trout, you'll want to be on the water from mid-April until mid-May. Kent Pond has a maximum depth of only 20 feet, not the best for sustaining season-long trout fishing. However, plenty of rainbow and brook trout are stocked during early season. Approximately 2,500 rainbows and 1,000 brook trout are stocked annually in the pond. A

Kent Pond on the east side. A formal VF&WD boat launch is located on the opposite side of the pond. Small craft can also be launched from the pond's east side, off Thundering Brook Road.

number of 2-year-old brook trout up to 14 inches and 2-year-old rainbows up to 16 inches are in the mix.

The stockings are planted on both the east and west sides of the pond. During May you'll likely encounter conventional tackle anglers fishing along Thundering Brook Road at the fishing platform next to the dam on the east side of the pond. It's possible to launch a canoe or kayak from that side of the pond at a small launching area along Thundering Brook Road. Parking isn't as expansive as that found at the VF&WD access site on the west side. A parking area is also located near the fishing platform.

Deerfield River

A popular trout stream that spans two states, the Deerfield River begins its 70-mile journey from springheads in Vermont's expansive Green Mountain National Forest. The East Branch of Deerfield's headwaters near the Stratton area gathers Black Brook then traverses south, unloading into Somerset Reservoir. A 5-mile-long section of the East Branch discharges as a tailrace from Somerset Reservoir, one of several impoundments the river enters along its length. After flowing through a remote area popular with trail hikers and other winter sport recreational users, the East Branch combines water with the main Deerfield River. Both inflows contribute their loads to the 30-acre Searsburg Reservoir.

The Deerfield River pushes on a few miles, enters Harriman Reservoir, then resumes its journey, discharging as a tailrace from Harriman Dam. After flowing several miles, it departs Vermont and transfers its water to Massachusetts at Sherman Reservoir. After exiting Sherman Reservoir near Readsboro, the river continues to Fife Brook Dam at the town of Florida. Its tailrace below Fife continues south in a narrow valley, easterly through Charlemont, Massachusetts, and then through Shelburne Falls. After flowing through

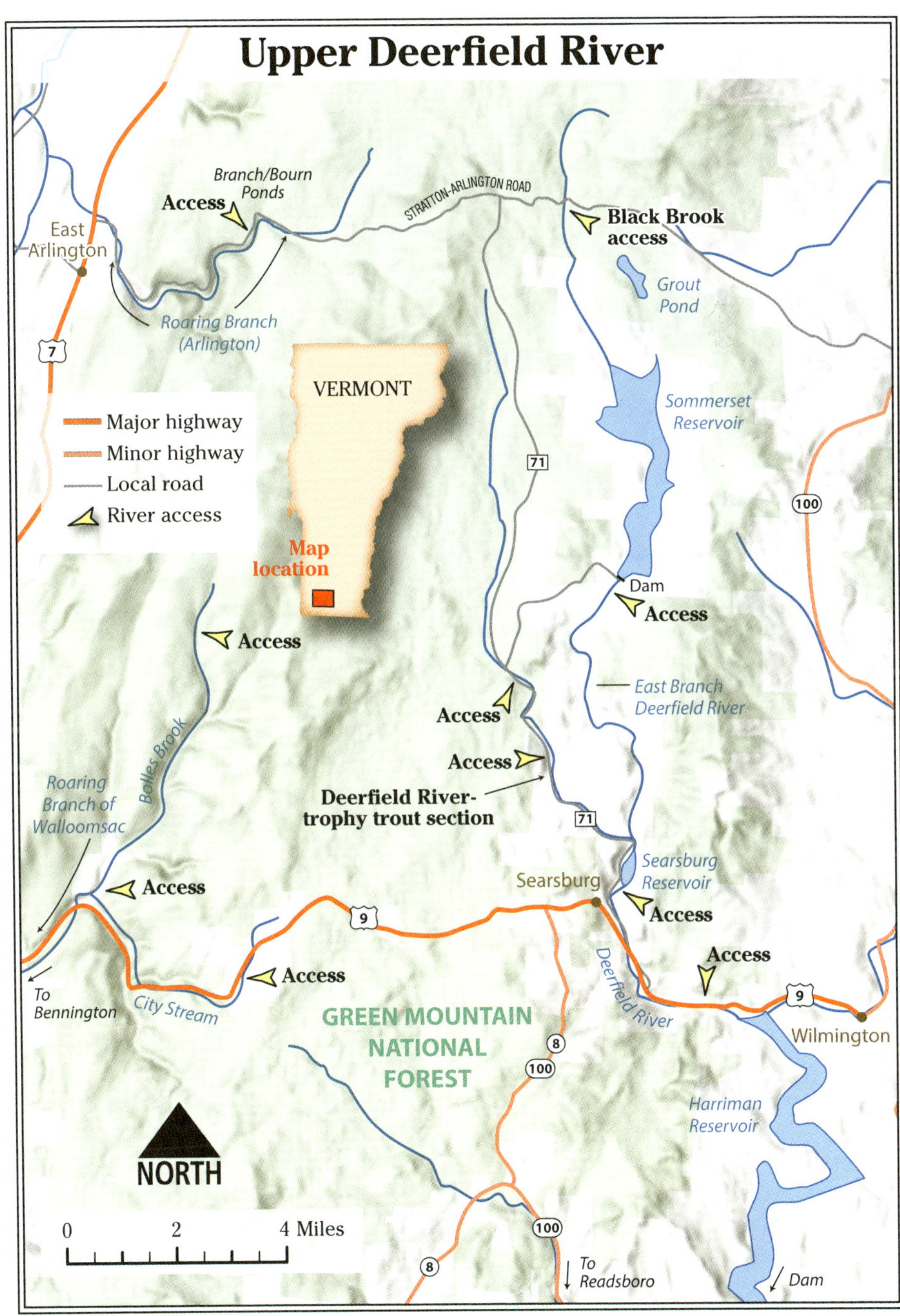

Upper Deerfield River
Branch/Bourn Ponds
Access
STRATTON-ARLINGTON ROAD
Black Brook access
East Arlington
Grout Pond
Roaring Branch (Arlington)
7
VERMONT
Sommerset Reservoir
Major highway
Minor highway
Local road
River access
71
100
Map location
Dam
Access
Access
East Branch Deerfield River
Access
Bolles Brook
Access
Roaring Branch of Walloomsac
Deerfield River-trophy trout section
71
Searsburg Reservoir
Searsburg
Access
Access
Access
9
Access
To Bennington
City Stream
Deerfield River
9
Wilmington
GREEN MOUNTAIN NATIONAL FOREST
8
100
Harriman Reservoir
NORTH
0
2
4 Miles
100
8
To Readsboro
Dam

Deerfield River's bypass section along VT 9, Molly Stark Byway, downstream from the hydroelectric plant. Fast-moving rapids and pocketwater are perfect rainbow trout habitat. VF&WD has evaluated stocking different strains of rainbows in this section, posting signage to alert anglers of the study. VALERIE VALLA

Shelburne and under the historic Bridge of Flowers, the Deerfield flows toward its confluence with the Connecticut River in Greenfield, Massachusetts.

While many dams dot both the Vermont and Massachusetts Deerfield River watersheds—there are at least 33 in Vermont and more than 50 in Massachusetts—the number of major dams along the Deerfield River itself is striking. With Harriman in Vermont being the first, constructed in the early 1920s, and Fife being the last, constructed in the early 1970s, the character and mood of the river changed forever.

SEARSBURG RESERVOIR DAM SPILLWAY TO HARRIMAN RESERVOIR: THE SEARSBURG BYPASS

The "bypass" mileage between the Searsburg Reservoir dam spillway and the next impoundment that swallows its flow, Harriman Reservoir, is one of the more popular fly-fishing sections along Vermont's upper Deerfield. There's plenty of access along the stretch, which runs for some 4 miles. If approaching from the south, you'll want to drive to the VT 9/Somerset Road intersection, approximately 5 miles northwest of Wilmington and approximately 14 miles east of Bennington along VT 9. Somerset Road is also designated as FR 71, a gravel road often ungraded, with plenty of ruts and potholes along the way. The gravel road is also narrow, so be alert for oncoming vehicles. Park at the lot next to the dam, approximately 1 mile up FR 71. You'll note the small white dam operations building along the dam spillway next to the access lot used by anglers and others.

A short walk from the parking area will take you over an interesting 8-foot-diameter wood-stave penstock conduit that runs along the river. Part of the discharge from the Searsburg Reservoir dam passes through the penstock then rejoins the river 3.5 miles downstream, where it ends its journey at a small hydroelectric facility situated on the riverbank

The striking 3.5-mile-long wood-stave penstock conduit begins at Searsburg Dam. A portion of water discharged at Searsburg Dam feeds a small hydroelectric facility located a few miles downriver near Wilmington. VALERIE VALLA

near the Wilmington town line. The river itself continues from the dam's spillway and pushes downstream to the town of Wilmington, where it joins discharged penstock water.

Continue walking past the penstock toward the river and you'll reach a path and short set of wood stairs, also used by recreational watercraft enthusiasts, that leads to the water. Take special note of the posted warning signs. Water levels can increase at any time, and it's important to stay alert. Depending on the season, even the day, water here can run from high and not safe to wade to lower flow levels quite manageable to fish. Flows during late season can be very low, as is true with most free-flowing rivers that time of year. For many years up until the late 1990s, when the Federal Energy Regulatory Commission (FERC) mandated water flow standards through their relicensing mandates, there were times when the Deerfield in this area and downstream often appeared as a lifeless, almost-dried-up streambed. That all changed, for the better of the fishery and fly-fishing opportunities.

Pay attention to warning signs that are posted along the river, alerting anglers and other users to keep vigilant for sudden water level increases. Signage recommends not wading too far out into the river. A useful online resource for determining water discharges exiting not only Searsburg Reservoir but also other impoundments along the course of the Deerfield River is Waterline: www.h2oline.com/default.aspx?pg=st&op=VT.

You'll want to sample the water here with a variety of fly patterns, depending on water levels. Much like the entire bypass section, streamer patterns will entice rainbows that

Harriman Reservoir at Wilmington is heavily stocked with brown, brook, and rainbow trout and some landlocked salmon.

are stocked annually. Caddis patterns, both dry flies and pupae, can bring fish to your net. During periods of moderate water levels, I've taken fish on Haystack and Wulff dry flies. During early-season angling, try patterns such as Black Leeches.

Other access areas along the bypass are located a couple of miles downstream from Searsburg Dam along VT 9 (Molly Stark Byway). From the VT 9/Somerset Road intersection, drive a little over 2 miles east to a section along the river next to the road where parking and access is very easy (about 3 miles west of Wilmington if approaching from that direction). You'll notice the small hydroelectric facility across the river where penstock water ends its journey and discharges into the river. The river is a moderate gradient. Plenty of rapids, pocketwater, and glides—perfect for rainbow trout—typify this section, depending on reservoir discharge levels. VF&WD signage along this river section alerts anglers to a study that evaluated two different strains of rainbow trout—Eagle Lake and Erwin-Arlee—for survivability, growth, and catchability.

The lower flats reach as it approaches the Harriman Reservoir can be a good place to spot trout rising to small *Baetis* flies and caddis. Harriman Reservoir is typically stocked with a few thousand brown, rainbow, and brook trout and a few hundred landlocked salmon. You might run into fall-run browns and landlocks in that area.

READSBORO SECTION AND WEST BRANCH OF DEERFIELD RIVER

Exiting Harriman Dam, at the southwest corner of the reservoir, the river pushes southward another 3 miles or so through isolated sections, where it eventually accepts inflow from the West Branch of the Deerfield in the town of Readsboro's crossroads. Flowing through steep terrain in a ravine setting, much of the Readsboro section is difficult to access. One option is to fish the river along the stretch just downstream from the Harriman Dam discharge, although getting into the water here can present challenges. Park at the dam picnic area

Harriman Reservoir's tailrace exits the dam then traverses through a ravine-like land feature, flowing some 3 miles to the town of Readsboro crossroads before entering Massachusetts. VALERIE VALLA

that's nicely arranged with tables and grills. The parking lot is popular with cross-country skiers, snowmobilers, hikers, and anglers. It's easily reached by driving 8.7 miles south on VT 100 from the intersection with VT 9 in Wilmington. Turn right on Dam Road just beyond the Harriman Reservoir Whitingham Car Top Carry-In Access. The road sign is small, so watch for it. The 1.7-mile-long gravel road will get you to the dam parking lot.

While it's a challenge to get into the river stretches directly below the dam, a couple of access options are possible, provided water levels are safe and conducive to wade-fishing. I can't overemphasize wading safety downstream from the dam. Don't wade too deep, and always be mindful of an escape route should water discharges rise unexpectedly. Always check Waterline for updated information on scheduled reservoir discharges. If you're not familiar with how to access tailwater areas of this nature safely, don't do it. Or locate an experienced fly angler or guide who knows how to safely fish this reach and can safely fish the area with you.

Directly to the left of the picnic area, you'll find a popular Catamount X-C Ski Trail, marked by small blue signs. Trek about 5 minutes down the trail to a junction with another trail that leads to the base of the dam, marked by directional signage. A sharp right at the junction will take you down a slope. It's a bit cumbersome, but just before you reach a small footbridge, you can reasonably cut through the woods to the east side of the river, which can be sampled, depending on water levels, for at least a short distance before running into steep banks.

Before reaching the footbridge, you'll notice another Catamount X-C trail (next to a row of timbers under the power line) leading to the left. This trail (an old railroad bed) won't get you anywhere near the river. The problem is that the trail parallels the river along a precipice so steep there's no practical way to get to the water. Another option is to continue walking past the footbridge to the base of the dam and the west side of the river. Depending on water levels, you can hang close to a steep slope, negotiate lots of boulders and rocks, and get to a few stretches downstream. This isn't always an easy task either; I actually favor other access points downriver instead of getting into the dam sections.

You can access the river by driving another 4 miles past the Dam Road/VT 100 intersection into the town of Readsboro. You'll cross over the high bridge in the center of the crossroads. Take an immediate right turn onto Phelps Lane. It will take some effort to get into the river, but you can park along the road shoulder and find the best way down a slope to get into water.

The river departs Vermont a mile or so downstream from the VT 100 bridge crossing in Readsboro, flowing into Sherman Reservoir then through one of the most scenic regions of Massachusetts. Before exiting into Massachusetts, the river flows along Tunnel Street. While you'll have to carefully trek down a somewhat steep hill to get into the water along that section, it's worth sampling before it quickly weds Sherman Reservoir near the Harriman Station Powerhouse. I like to fish this section during the fall season. Parking can be difficult, but there are a couple of areas where you can pull over once Tunnel Street departs a short stretch congested with dwellings and other small buildings. One parking area that can provide easy access is along the river side of the road shoulder, directly next to a small generator and "Welcome to Readsboro" sign. You can get into the river by making your way down the slight slope to the water.

Fly-fishing guide Brian Comfort, who operates Deerfield Fly Shop in South Deerfield (https://deerfieldflyshop.com), is undoubtedly one of the most enthusiastic Deerfield River anglers. While most of his Deerfield fishing occurs in Massachusetts, Brian has also sampled the Readsboro stretch. "You'll want a supply of Rich Strolis–tied streamers, pheasant tail nymphs, caddis patterns, and other flies you might typically cast on similar waters while fishing that

The town of Readsboro, off VT 100, is the lowermost Vermont section before the Deerfield River enters Massachusetts.

The Deerfield River at Readsboro, Vermont, not far upstream from the Massachusetts state line. Discharging from Harriman Reservoir some 4 miles upstream at the reservoir's dam, the Deerfield flows through an isolated section, eventually showing itself in the Readsboro crossroads area, where it picks up the West Branch of Deerfield River a short distance upstream from the reach shown here. During late season, such as early October, the river can run clear and shallow.

West Branch of Deerfield River, a short distance upstream from its confluence with the Deerfield River in the main settlement area of the town of Readsboro. Access this reach along VT 100, across from the Readsboro Lions Club Family Park. VALERIE VALLA

stretch," Brian said during a wonderful chat we enjoyed at his shop on a winter afternoon. "I'd recommend having sink-tip lines on hand, varying in length depending on the water levels." Brian scanned the shop bins and selected some of Rich Strolis's patterns. Rich's Black Avenger is a good choice.

If in the area, it's worth trying the West Branch of Deerfield River. This high-gradient stream tumbles along VT 100 until it dumps into the Deerfield River in the center of Readsboro crossroads. You'll need your wading staff for sure, jockeying around lots of boulders and rocks. I performed a DNA sample of the stream section near the Lion's Club Family Park. Both brook and brown trout DNA showed up in the analysis. During late season and early fall, the West Branch's water levels can be quite low yet still require careful wading.

HEADWATERS: MAIN DEERFIELD, BLACK BROOK TRIBUTARY, AND UPPER EAST BRANCH

A few interesting areas worth sampling exist in Deerfield River's headwaters, upstream from Searsburg Reservoir. Driving 1 mile north on FR 71 from Searsburg Reservoir Dam will get you to a steel footbridge that crosses the main Deerfield River upstream of the reservoir. While the river section here is generally unfavorable to self-sustain a trout population, VF&WD recently designated the section of river between the steel footbridge upstream approximately 4 miles as a Special Regulations section. Trophy sections are open from the second Saturday in April until October 31.

The river upstream from Searsburg Reservoir maintains natural stream flow through lands owned by the USDA Forest Service and Great River Hydro power company. In correspondence, USDA Green Mountain National Forest biological technician Scott Wixom told me, "The main stem of the Deerfield becomes pretty warm during the summer, resulting

The main Deerfield River a couple of miles upstream from Searsburg Reservoir Dam. The river in this area warms and becomes shallow for much of late spring into summer and can't sustain wild trout. Stocked annually with several hundred large 2-year-old brook trout, this reach is designated as a Special Regulations section. The triploid brook trout can't reproduce. VALERIE VALLA

in limited use by wild brook trout, although many of the tributaries are cool enough to support wild native populations." VF&WD elected to manage the trophy section as a "put-and-take" stream reach by stocking it with large 2-year-old brook trout raised at both the nearby Bennington Fish Hatchery and the Dwight D. Eisenhower National Fish Hatchery in North Chittendon, Vermont.

BLACK BROOK TRIBUTARY AND UPPER EAST BRANCH OF DEERFIELD RIVER

If you're interested in targeting wild brook trout in unstocked water, head for the East Branch of Deerfield River and Black Brook, a small tributary. Unless you're up for a long trek to lower stretches of the East Branch of the Deerfield via the East Branch Trail, accessed by the steel footbridge crossing, you're better off continuing north on FR 71 about 6.3 miles to Somerset Road to the East Branch's access point at the Somerset Reservoir dam. Along the way to the dam, just short of 4 miles, you'll come to the FR 71/Somerset Road intersection signage. Continue straight on Somerset Road for 2.6 miles to the dam and the East Branch's access point.

The trail, marked with blue Catamount Trail Cross Country Ski markers and orange boater portage markers, will deliver you to the stream. The trail splits a few hundred feet down the hill. Take the right-hand trail, marked with orange markers. You'll cross a small wood plank bridge that will get you into the water.

Cold-water discharges from the reservoir support a wild brook trout population. In the late fall of 2023, I performed a DNA sample with my handy Smith-Root device, running 2 liters of stream water through a 5-μm DNA filter to determine any presence of brown trout. Not surprisingly at all, results after analysis indicated only brook trout DNA.

East Branch of Deerfield River a short distance downstream from the Somerset Reservoir dam. Water here is ice cold during the heat of summer, perfect for the wild brook trout that inhabit this stream reach.

If interested in sampling Black Brook tributary and uppermost East Branch above Somerset Reservoir, continue north on FR 71 at the FR 71/Somerset Road intersection you passed on the way to the Somerset Reservoir dam. Drive a bit under 9 miles to Stratton-Arlington Road. A right turn will almost immediately cross Black Brook. The small stream is located next to a large parking area at the Stratton Mountain Pond trailhead. The brook flows through a wide-open area north of Stratton-Arlington Road at the parking area but then quickly vanishes downstream of the small bridge crossing. (The 3-mile trek up the trail to 46-acre Stratton Pond, a brook trout pond, is a great fall-of-the-year adventure. Primitive camping is allowed at the pond.)

I prefer stretches downstream from the bridge, where the stream better presents its possibilities. Keep in mind that during late summer, this reach can dry up to a trickle. Brook trout are there, often hidden under cover of rocks. Fishable small pools can be few and far between. Small bead-head flies plopped in pocketwater pools will interest the stream's wild brook trout. The typically small brook trout also like a Lime Sally Stonefly.

One strategy, if you're into taking a hike, is to park at the Stratton Mountain Pond trailhead parking lot then walk back to FR 71. A walk down the road approximately 0.6 mile will intersect the Appalachian Trail. Enter the trail on the right side of the road (Appalachian Trail signage is only located on the left side of the road, so keep your eyes open for the trail on the right side). Trek a good distance on the trail to a footbridge crossing that will get you into the brook. Then fish upstream, casting small dry flies through the pocketwater, working your way back to the parking area. The National Geographic Green Mountain National Forest South map 748 clearly shows the parking areas, roads, and trails.

Access to the upper Deerfield River is in the same neighborhood. Like Black Brook, by late summer the water levels can dry up to a trickle. Drive 1 mile or so east of the Stratton Mountain Pond parking area to the Stratton Mountain trailhead lot. You'll pass over the

Black Brook headwaters at the Stratton Pond Trailhead parking area on Stratton-Arlington Road. The brook departs the open area then quickly vanishes into tree cover as it flows south through small riffles, runs, and pools that hold wild brook trout.

Ausable Wulff *Tied by Fran Betters, Adirondack Sport Shop*

Schmidtmann-Valla Smelt *Tied by Mike Valla*

little stream just before reaching the parking area. Bring your 7-foot, 3-weight rod for fishing Black Brook and the uppermost sections of the Deerfield River.

If you're approaching the upper sections of these two streams from the west out of the Arlington, Vermont, area, Kelly Stand Road will get you to Stratton-Arlington Road, as will Mountain Road near the Stratton Mountain Ski Resort area if approaching from the north. If in the area, you'll want to check out three interesting ponds—Branch Pond, Bourn Pond (brook trout ponds), and Grout Pond (a warmwater fishery). Bourn and Branch Ponds are considered part of the nearby Hudson River Drainage Basin.

West River

West River, a major southern Vermont Connecticut River tributary, flows nearly 54 miles from its source tributaries in the Green Mountain National Forest near Mount Holly in southeastern Rutland County. It's been written that its historical name was *Wantastiquet* ("Waters of the Lonely River"). Of note, streams like Mill River that originate on the northern aspect of Mount Holly flow toward Otter Creek and into Lake Champlain.

West River at Depot Street Bridge, just downstream from Jamaica State Park. Part of the stocked section on the river, this reach generally experiences periods of elevated water temperatures during summer. It's best to sample the area in mid-spring.

On its way to Brattleboro, where it joins the Connecticut River, the West River traverses southerly, passing through Weston, Londonderry, Jamaica, Townshend, and a couple other small towns along its path. Its flow is interrupted by a couple of US Army Corps of Engineers dams that create Ball Mountain and Townshend Lakes. Recreational whitewater floaters enjoy negotiating rapids between the two lakes twice a year when planned water releases from the upstream dam occur, usually in April and again in September. In summer swimmers gather at "The Salmon Hole," located in 772-acre Jamaica State Park.

During May, fly anglers enjoy annual rainbow trout stockings that take place downstream from Ball Mountain Reservoir at the state park. Some 800 rainbows are stocked at the park area, downstream to the Depot Street bridge in Jamaica. The park provides tent/RV sites for those wishing to enjoy an added experience. A railroad operated along the riverbank at what is now the park from 1879 to 1923 but was wiped out by a flood. Today the old railbed serves as a 2.5-mile-long trail that leads to the dam from the park. A July 2023 flood destroyed part of the trail, and it was off-limits at the time of this writing. Since the trail is popular with visitors, the damaged riverside trail will undoubtedly be addressed and repaired.

In addition to opportunities at Jamaica State Park, fly fishers interested in headwater experiences can find those well upstream from Jamaica at locations north of Weston. Two headwater tributaries that have been called "gems" can provide fishing for stream-bred brook trout.

Jenny Coolidge Brook and Greendale Brook

While the West River warms significantly during summer months, anglers can find abundant brook trout in two of its tributaries upstream from Weston. Reach both brooks via a 5-mile drive from Weston. Take VT 100 in Weston a couple of miles to a left on Greendale

Greendale Brook, a West River tributary, near Glendale Campground in the Green Mountain National Forest. Stream-bred brook trout inhabit the small pools and runs along the tributary's upper reaches. VALERIE VALLA

Jenny Coolidge Brook at the Greendale Road crossing, a short distance upstream from its confluence with Greendale Brook. Water levels can drop during summers that experience low rainfall, but temperatures usually remain good for the small brook trout that swim in its small pools.

Road (FR 18), where you'll cross a headwater section of the West River. Continue on Greendale Road as it follows its namesake brook upstream 2.1 miles to the Greendale Road/Jenny Coolidge Road (FR 17) intersection and Jenny Coolidge Brook bridge crossing.

You'll encounter a couple of road shoulder parking turnouts along Glendale Road for access into Greendale Brook as you make your way upstream toward Jenny Coolidge Brook. Jenny Coolidge Brook joins Greendale Brook a short distance downstream from the bridge crossing. It's best to park your vehicle at the large parking area next to the bridge, take a trek up FR 17 (Jenny Coolidge Road) a distance, and then fish downstream through the pockets and pools back to the parking area. Small wild brook trout and a few brown trout inhabit the pools. You can also drive along FR 17 (depending on the time of year) and follow the brook quite far upstream. The road is gravel/dirt and narrow in areas, and you might run into difficulty should you find it necessary to turn around and drive back down the mountain.

I once drove way up the road to sample the brook's uppermost stretches, only to be greeted by a fallen tree that blocked me from continuing to a turnaround area. I had to back down the hill with some difficulty—and angst. Keep in mind that Green Mountain forest roads are often closed in early spring through "mud season." My preference for fishing both Jenny Coolidge Brook and the headwaters of Greendale Brook is during late summer. However, once things dry out in mid-spring, both streams are nice to fish. During late summer in some low-rainfall years, both Jenny Coolidge and Greendale Brooks can experience very low water levels, although water temperatures remain cold enough for ethical fishing. I usually concentrate my fishing on the uppermost reaches.

Greendale Brook flows through a heavily shaded area starting 1.2 miles upstream from the Greendale Road/Jenny Coolidge Road intersection bridge crossing. In 1 mile you'll encounter the nice yet small Greendale Campground, operated by Green Mountain National Forest. Each of its 11 campsites, situated directly next to the brook, has a picnic table. A privy facility is on-site. The campsite fee is by self-payment at the site; site #1 is particularly nice. Greendale Road continues another 0.2 mile beyond the campground, where it dead-ends at a bridge crossing. A small feeder tributary enters the brook just downstream from the bridge. Years ago, the area upstream from the bridge, now a wide-open area, was occupied by a large beaver dam that backed up the water. The dam has long since been breached, and the once-flooded area is reverting to a beaver meadow of sorts, with saplings and growth. You'll want to fish downstream from the bridge.

One strategy is to start at the defunct bridge and fish downstream back to the campground, twitching small dry flies through the pools. I usually park at the campground, fish downstream to a bridge crossing point where I can easily get back up to the road, then walk the road back to the campground. You'll likely pick up small wild brook trout in the small pools.

Williams River

If I had written this Williams River section prior to late August 2011, or during the late 1980s, you'd be reading a much different story about where to fly fish—the best stretches and reaches and other details that might inspire those who have been considering casting on its waters. To say the Williams has changed since Tropical Storm Irene's wrath in 2011 would be an understatement. Of course, many other streams throughout Vermont and the

Williams River headwaters along Smokeshire Road. The upper reaches of the Williams River are clear and cold during late summer. VALERIE VALLA

Northeast were ripped open by the storm, its heavy rains and floods altering what were once good fly-fishing reaches. Whenever my travels take me through Chester on VT 11, I shake my head as I drive along the Middle Branch of Williams River, viewing washed-out riverbanks along the way. Subsequent storms also have opened wounds along the stream. Yet that said, the Williams deserves at least some mention.

A nearly 30-mile-long tributary of the Connecticut River, the Williams flows south of the Black River and north of Saxtons River. Along its path, which began near Andover, the stream picks up water from the Middle Branch of Williams River at Chester; the South Branch of Middle River adds its flow just southeast of Chester. Once the Williams meets Chester, it flows close to VT 103, passing under Bartonsville Covered Bridge in Rockingham, then soon Worrall Covered Bridge before the river meets the Brockways Mills hydroelectric area and crashes through Brockways Mills Gorge. In another 5 miles the river enters the Connecticut at Herricks Cove.

In the years leading up to 2011, just prior to Tropical Storm Irene, VF&WD used to stock the lower reaches with a small number of rainbow trout. The upper Williams River and Middle Branch reaches received a small number of brook trout. Up to the late 1980s, about the time we moved to eastern New York state not all that far from Vermont, larger numbers of rainbow, brook, and brown trout were stocked. Today, no stocking occurs throughout the river system. Much of the river structure upstream from Brockway Mills Road has been lost to the storm, degraded to the point that habitat was virtually destroyed in some stretches. However, small brook trout, mostly young-of-the-year but a few dollar bill–size, inhabit the headwaters of Williams River system.

South Branch of Williams River along Popple Dungeon Road, 3 miles upstream from the VT 35/Popple Dungeon Road intersection.

SOUTH BRANCH WILLIAMS RIVER HEADWATERS

In Vermont, VF&WD biologists have a long-term survey site on the South Branch Williams River along Popple Dungeon Road, several miles upstream from the VT 35/Popple Dungeon Road intersection. In 2021 biologists estimated 185.26 young-of-the-year brook trout per mile. Upstream reaches have good forest cover, and the water is cold during late summer. The reach in that area is small. I used a 6-foot, 3-weight rod the last time I fished the upper reach that flows next to Williams River State Forest. Size 18 barbless Lime Sally dry flies cast along the small pools resulted in only small young-of-the-year brook trout, with only one french fry–size fish. While no sizable trout were caught, I had a blast wet-wading the stream during a particularly hot August day. Park at road shoulder turnouts along Popple Dungeon Road for access to its upper reaches.

WILLIAMS RIVER HEADWATERS

Sections along the Williams River upstream from Chester near Wymans Falls Road don't appear as devastated as the lower reaches. Some very nice pools and structure can be found on other stretches upstream from Jewett Road. I didn't have time to fish all those sections in 2024, but I took DNA samples at Jewett Road crossing and one farther upstream along Smokeshire Road, where the small stream flows through a very nice area well shaded by hemlock overstory. Not surprisingly, results at the Jewett Road crossing showed no brook trout DNA flowing through and only a small presence of brown trout DNA.

On a hot, late-August day in 2024, the water upstream along Smokeshire was cold as I wet-waded through the nice pools and runs. One small brook trout ate a small Lime Sally dry fly. I took a DNA sample at the site and was surprised that brown trout DNA numbers

The mouth of the Williams River at Herrick's Cove on the Connecticut River is a great place to latch into warmwater species.

came through, with 100 percent base-pair match, and only an extremely low number of brook trout DNA with 100 percent base-pair match.

WILLIAMS RIVER MOUTH AT CONNECTICUT RIVER

Given the Williams River's declining trout opportunities, fly angler Stan Steen, whom I ran into on the Black River during a May outing, prefers to head to the mouth of the Williams, targeting warmwater species. Head to Bellows Falls for easy access at the Great River Hydro recreation and picnic area on Herrick's Cove Road off US 5N. When you approach the area, turn left at the Great Hydro signage to get to the main boat launch. If you're fishing out of a canoe or other small watercraft, you can get into the cove by taking at right at the signage. Drive past the picnic area to the far end of the road. A narrow yet well-walked footpath leads to the water. You'll get into good smallmouth fishing in the cove.

Saxtons River

Just south of Williams River's mouth and Bellows Falls, the Saxtons River joins the Connecticut River. Its 22.9 miles begin as a trickle near Glebe Mountain in Windham. The stream then flows almost entirely along VT 121 as it heads to its confluence with the Connecticut River. The Saxtons flows as small water as it heads into Houghtonville. Just west of Houghtonville, the stream flows through a deep ravine. I've crawled down into the ravine and stream at a bridge crossing about 7 miles upstream from Houghtonville; it's nice water, but I wasn't able to get a fish. Access along the forested stream reach is provided as a courtesy to the public by Meadowsend Timberlands LTD. Small yellow signs are posted on trees along VT 121 in that area. As long as visitors respect their lands for low recreational use, "leaving the land better than you found it," they'll continue to keep it open. (It was quite disconcerting when I discovered a large bag of trash tossed down the hill from the sign.)

Once past Grafton, the Saxtons River continues southeasterly, picking up the South Branch of Saxtons River just below Grafton. Winnie Park, a small streamside area with a small pavilion, is located on VT 121 in Grafton, just downstream from the VT 121/VT 35 intersection. The now-widened flow, with the added water volume, continues through Cambridgeport and then the village of Saxtons River, where it tumbles over a scenic waterfall before making its final approach into Connecticut River. Like the Williams River and others, Saxtons River's main stem trout fishery has experienced a decline over the last couple of decades. VF&WD regularly surveys the Saxtons River at a long-term monitoring site in Houghtonville, upstream from Grafton. VF&WD fisheries biologist Lael A. Will's July 2022 *Inland Fisheries* report revealed interesting Saxtons River data.

Trout population numbers between 2012 and into 2016 were much different seven years later, into 2022. In 2012, data revealed that the combined number of young-of-the-year and fish greater than 6 inches was well over 600 per mile. By 2022, numbers had dropped to almost none. "While the Saxtons River site historically had moderate trout populations, recent surveys indicate the population is essentially nonexistent. Warming temperatures and extreme drought and flood conditions that have occurred recently are likely contributing to these noted declines," Lael reported. It will be interesting what new data from subsequent population surveys might reveal. Stream systems have been known to heal and recover, especially through habitat improvement, should such activity be deemed an appropriate healer for the Saxtons River system.

Saxtons River at Winnie Park, a small riverside rest area in Grafton. This photo was taken in late August, during seasonal low-water conditions.

The upper Saxtons River along Meadowsend Timberlands forested lands. The small stream flows through a ravine, well shaded by overstory cover. This photo was taken during late August in very low-water summer conditions.

SOUTH BRANCH SAXTONS RIVER

While South Branch of Saxtons River is not what I'd call a significant fly-fishing destination, it's worth a look and cast or two if you're passing through the area or exploring any trout population changes that might have occurred since the last survey. The Windham Foundation access area is located less than a mile south of Grafton on VT 35. Park at the access area. A handsome footbridge leads to the opposite side of the stream, where a gravel path leads to McWilliam Covered Bridge, a small example of the covered bridges scattered throughout Vermont. The gravel path follows the small stream to Grafton Village Cheese Company, popular with visitors to the area. I took a DNA sample at this location on March 21, 2025. Not surprisingly, 100 percent base-pair matches for brook trout came through. However, I was surprised to see 100 percent base-pair matches for *Oncorhynchus*—rainbow trout.

South Branch Saxtons River at the Windham Foundation access point off VT 35, less than a mile south of Grafton. VF&WD biologists regularly survey the South Branch. VALERIE VALLA

Selected Fly Pattern Recipes

Al's Little Perch

Tied by Al Pitt

- **REAR HOOK:** #4 wet fly
- **FRONT HOOK:** #2 wet fly
- **THREAD:** White 6/0
- **CONNECTING WIRE:** Surfstrand 20-pound 7-strand silver
- **BODY:** Gold Holo Braid tinsel
- **BELLY:** Orange Slinky Fiber
- **UNDERWING:** Black-barred olive Voodoo Fibers or Grizzly Hair
- **OVERWING:** Black Slinky Fiber
- **EYES:** Painted yellow with a black pupil

Big Fish Fly

Tied by Ralph Graves

- **HOOK:** #8-10 Mustad 9671
- **THREAD:** Yellow Danville
- **TAIL:** Yellow calf tail
- **BODY:** Pinkish floss or dubbing; palmer with grizzly dyed yellow
- **WING:** Yellow calf tail
- **HACKLE:** Grizzly dyed yellow

Ausable Wulff

Tied by Fran Betters, Adirondack Sport Shop

- **HOOK:** #10-12 Mustad 94840 or 9671
- **THREAD:** Hot orange
- **TAIL:** Guard hair from woodchuck
- **BODY:** Australian opossum dyed rusty orange
- **WING:** White calf tail
- **HACKLE:** Grizzly and brown hackle

CDC Grannom Pupa

Tied by Tom Baltz

- **HOOK:** #12-16 Daiichi
- **THREAD:** Olive Danville Flymaster
- **BEAD:** Black tungsten
- **RIBBING:** Chartreuse UTC Ultra Wire
- **BODY:** Darker hair from hare's mask dyed chartreuse
- **BACK:** Black goose or turkey biot
- **SWIMMERS:** Two pheasant tail fibers on each side, treated with head lacquer
- **SPARKLERS:** Chartreuse Krystal Flash
- **HACKLE:** CDC, two turns
- **THORAX:** 50-50 blend gray squirrel and CDC clippings

Battenkill Flats

Tied by Tom Miller

- **HOOK:** #14-24 Mustad 94840
- **THREAD:** Brown Danville 6/0 waxed
- **TAIL:** Dun hackle fibers, splayed
- **BODY:** Stripped Rhode Island Red hackle stem
- **WING:** Dun rooster

Champlain Special Tandem

Tied by Scott Biron

- **HOOK:** #4-6 Mason Multibraid SS Wire
- **THREAD:** Of choice
- **BODY:** Yellow floss
- **RIB:** Flat gold tinsel
- **WING:** Sparse white bucktail with sparse yellow over and peacock herl
- **THROAT:** Red bucktail
- **HEAD:** Black thread

Cherry Peel Optic Bass

- **HOOK:** #2/0-1/0 Partridge Universal Predator
- **TAIL:** Red hen hackle barbs
- **THREAD:** Red 6/0 UNI-Thread
- **RIB:** Oval gold tinsel
- **BODY:** Cincinnati Red Lion Brand yarn
- **WING:** Red cashmere goat (bottom and top) and white cashmere goat (center)
- **THROAT:** Red hen hackle
- **HEAD:** Gloss yellow Testors enamel paint eye and gloss black enamel paint pupil

Copper Kettle

Tied by Al Pitt

- **REAR HOOK:** #4 wet fly
- **FRONT HOOK:** #2 wet fly
- **CONNECTING WIRE:** 20-pound Surfstrand
- **THREAD:** Fluorescent orange
- **BODY:** Copper Holo Braid
- **BELLY:** Orange Slinky Fiber
- **UNDERWING:** Copper Mirror Krinkle Flash
- **OVERWING:** Brown bucktail

Drew's Bugger

Tied by Drew Price

- **HOOK:** #2-6 Ahrex PR360 50 Degree Jig Hook
- **BEAD:** Hareline 3.8 mm Mottled Olive Insta Jig Tungsten
- **THREAD:** Tan Uni Thread 60
- **WEIGHT:** 10 wraps 0.025 lead wire
- **HACKLE:** Grizzly olive
- **RIB:** Medium copper wire
- **BODY:** Medium root beer Estaz
- **TAIL:** Olive grizzly marabou
- **FLASH:** Chartreuse Krystal Flash

Edson Tiger-light

Tied by Mike Valla

- **HOOK:** #6-8 streamer
- **THREAD:** Black 6/0 UNI-Thread
- **TAIL:** Barred wood duck flank
- **TAG:** Fine gold tinsel
- **BODY:** Peacock herl
- **WING:** Yellow bucktail; two small red hackle points on top

Female Rusty DNA Spinner

Tied by Henry Ramsay

- **HOOK:** #14 Daiichi 1180
- **THREAD:** Rust 8/0 UNI-Thread
- **TAIL:** Blue dun Microfibbets divided with a ball of pale-yellow dubbing
- **BODY:** Rusty brown goose biot
- **THORAX:** Rusty brown dubbing
- **WING:** Shrimp DNA Frosty Fish Fibers

Golden Darter

Tied by Mike Valla

- **HOOK:** #10-14 Mustad 3665A
- **THREAD:** Black 6/0 UNI-Thread
- **TAIL:** Slip of mottled turkey
- **BODY:** Yellow Danville floss, four-strand
- **RIB:** Flat gold tinsel
- **THROAT:** Tip of jungle cock body feather
- **WING:** White calf tail

Golden Stonefly Nymph

Tied by Al Pitt

- **HOOK:** #6 4X long
- **BEAD:** 3/16 gold
- **THREAD:** 6/0 Flymaster white
- **TAIL:** 2 amber goose boots
- **ABDOMEN:** Root Beer Pearl Chenille
- **THORAX:** Light grayish hare's ear dubbing

Note: This fly is based on a combination of Helleckson patterns.

Governor Aiken Tandem

Tied by Scott Biron

- **HOOK:** #4-6 Mason Multibraid SS Wire
- **THREAD:** Of choice
- **TAIL:** Barred wood duck
- **BODY:** Flat silver tinsel ribbed with oval silver tinsel, 3 red beads then 3 white beads on connector wire (some are tied without beads)
- **THROAT:** White bucktail
- **WING:** White and lavender bucktail, 6 strands of peacock herl
- **CHEEK:** Jungle cock eye
- **HEAD:** Black

Note: This is based on Jim Warner's pattern.

Hendrickson CDC Thorax Dun

Tied by Henry Ramsay

- **HOOK:** #14 Daiichi 1180
- **THREAD:** Pink 8/0 UNI-Thread
- **TAIL:** Blue dun Microfibbets divided with a ball of pinkish gray tan dubbing
- **BODY:** Pinkish gray tan goose biot
- **THORAX:** Pinkish gray tan dubbing
- **WING:** Natural dun CDC tied as a post
- **HACKLE:** Brown dun

Juggernaut 2.0

Tied by Rich Strolis

- **REAR HOOK:** #1-4 Ahrex TP 610
- **THREAD:** Olive Veevus 140 power thread
- **TAIL:** Olive Whiting American Rooster Saddle
- **BODY:** Olive EP Sparkle Brush
- **WING:** Olive Arctic fox tail reverse tied
- **BEAD/CONNECTION:** Olive 3D Pro bead and 19 strand Beadalon
- **FRONT HOOK:** #1/0-2 Ahrex TP 610
- **REAR COLLAR:** Olive marabou palmered
- **BODY:** Olive EP Sparkle Brush
- **WING:** Olive Arctic fox tail
- **COLLAR:** 2 bunches of olive Senyo Laser dub or Letera Magnum dub
- **HEAD:** Green Fish Skull Bandito Baitfish Head #5

Lime Sally

Tied by Mike Valla

- **HOOK:** #16-18 Daiichi 1180 or Mustad 94833
- **THREAD:** Chartreuse 6/0 Danville
- **BODY:** Chartreuse Fly-Rite dubbing
- **THORAX:** Chartreuse Fly-Rite dubbing
- **WING:** Medium gray hackle tips or medium gray poly yarn
- **HACKLE:** Medium dun palmered over thorax

Little Brook Trout

Tied by Mike Valla

- **HOOK:** #4-6 Mustad 3665A
- **THREAD:** Black 6/0 UNI-Thread
- **TAIL:** Green bucktail
- **BODY:** Cream fox fur
- **RIB:** Silver flat tinsel
- **THROAT:** Orange bucktail
- **WING:** White bucktail, orange bucktail, and badger hair
- **CHEEKS:** Jungle cock eyes

Maple Syrup

Tied by Al Pitt

- **HOOK:** #10 1X long
- **BEAD:** 1/8 gold
- **TAIL:** Red or yellow marabou
- **BODY:** Light tan chenille

Masked Avenger

Tied by Rich Strolis

- **REAR HOOK:** #1-4 Ahrex TP 610
- **FRONT HOOK:** #1/0-2 Ahrex TP 610
- **THREAD:** Black Veevus 140 Power Thread
- **TAIL:** Black marabou
- **BODY:** Purple Polar Reflector Flash
- **WING:** Arctic fox tail reverse tied in the round, black with purple grizzly saddles on each side
- **BEAD:** 3D Pro bead and 19 strand purple Beadalon

- **BODY:** Purple Polar Reflector Flash; 3–4 turns black marabou
- **WING:** Arctic fox tail and purple grizzly saddles
- **COLLAR:** 3–4 turns EP black Foxy Brush topped with 5–6 strands of black Flashabou
- **HEAD:** Fishmask
- **EYES:** Red oblong pupil

Masked Avenger Single

Tied by Rich Strolis

- **HOOK:** #1-4 Ahrex TP 610
- **THREAD:** Black Veevus 6/0 Power Thread
- **TAIL:** Black Bugger marabou
- **BODY:** Black Veevus UV Body Fuzz (large)
- **WING:** Black Arctic fox body
- **COLLAR:** Midnight Senyo Fusion Dub

Montreal White Belly Tandem

Tied by Scott Biron

- **HOOK:** #4-6 Mason Multibraid SS Wire
- **THREAD:** Of choice
- **TAIL:** 2 golden pheasant crests
- **BODY:** Embossed silver tinsel
- **THROAT:** Red bucktail sparse
- **WING:** White under red bucktail with 8 peacock herls over
- **CHEEK:** Jungle cock eyes
- **HEAD:** Black thread

Mr. Bow Jiggles

Tied by Drew Price

- **HOOK:** #2/0-2 Gamakatsu Light Wire Live Bait Hook
- **THREAD:** Danville 210 denier white, colored to match fly with marker
- **EYES:** Large (5.5 mm) lead or tungsten dumbbell
- **WEIGHT:** 10 wraps of 0.035 lead wire coated in super glue
- **ANTENNAE:** 2 strands of barred and speckled crazy legs in orange or olive
- **CLAWS:** Pellet FNF Slush Jelly (singed at the end with a lighter to seal it)
- **BODY:** Large Badger Flexi Squishnille UV Olive, UV Rusty Brown or Black Legged Crawfish Orange Core
- **HEAD CEMENT:** Thin UV Solarez

MV Cossayuna Fathead Minnow

Tied by Mike Valla

- **HOOK:** #6-10 Mustad 79580
- **THREAD:** Tan 6/0 UNI-Thread and black 6/0 UNI-Thread
- **TAIL:** Olive cashmere goat
- **RIB:** Silver Lagartun flat French tinsel (medium)
- **BODY:** Lion Brand Fishermen's Wool Yarn (#98)
- **WING:** Brown bucktail dyed yellow and brown bucktail dyed olive, with IDM4 Blue Ice Dub Shimmer Fringe fibers
- **EYE:** Gloss yellow Testors enamel paint iris and gloss black enamel paint pupil

MV Emerald Shiner

Tied by Mike Valla

- **HOOK:** #4-8 Mustad 3665A
- **THREAD:** White 6/0 UNI-Thread and olive 6/0 UNI-Thread
- **TAIL:** Dun hen hackle barbs
- **RIB:** Silver Lagartun flat French tinsel (medium)
- **BODY:** Lion Brand Fishermen's Wool Yarn (#98)
- **THROAT:** White bucktail
- **BODY FLASH:** Peacock and pink Hareline Ice Dub Shimmer Fringe
- **WING:** Brown bucktail mixed with olive brown bucktail
- **HEAD:** Gloss white Testors enamel paint eye and gloss black enamel paint pupil

Pheasant Wing Hexagenia

Tied by Mike Valla

- **HOOK:** #8-10 Mustad 9671
- **THREAD:** Tan 8/0 UNI-Thread
- **TAIL:** 2 porcupine quills
- **BODY:** Light brown dubbing of choice; yellow dubbing thorax
- **RIB:** Yellow floss
- **WING:** 2 pheasant body feathers
- **HACKLE:** Grizzly

Polar Sculpin

Tied by Ken Tutalo

- **HOOK:** Mustad saltwater hook of choice; tandem attachment: 20-pound fly line backing, orange or color of choice doubled over, size of choice
- **THREAD:** Of choice
- **TAIL:** Olive marabou
- **BODY:** UV Polar Chenille
- **COLLAR:** Olive brown EP Streamer Brush
- **HEAD:** Olive sculpin helmet

Red Quill DOA Cripple Emerger

Tied by Henry Ramsay

- **HOOK:** #14 Daiichi 1130
- **THREAD:** Rust 8/0 UNI-Thread
- **BODY:** Rusty brown goose biot
- **THORAX:** Rusty brown dubbing
- **SHUCK:** Brown Zelon fibers
- **LEGS:** Rusty brown CDC fibers
- **WING:** Natural dun CDC

Sandbar Smelt Tandem

Tied by Scott Biron

- **HOOK:** #4-6 Mason Multibraid SS Wire
- **THREAD:** Of choice
- **BODY:** Orange floss
- **RIB:** Flat silver tinsel
- **BELLY:** Short yellow calf tail with white bucktail
- **THROAT:** White hen with yellow hen
- **UNDERWING:** 4–6 peacock herls
- **WING:** 2 white saddles with 1 yellow saddle slightly shorter on the outside

Schmidtmann-Valla Smelt

Tied by Mike Valla

- **HOOK:** #2 Tiemco 9395 or Mustad 9674
- **THREAD:** Olive 6/0 UNI-Thread
- **TAIL:** Olive hen hackle tips
- **UNDERBODY:** Lion Brand Fishermen's Wool Yarn (#098)
- **BODY:** Mylar tubing (medium)
- **THROAT:** Dun hen hackle and olive hen hackle
- **WING:** Olive cashmere goat over black cashmere goat
- **EYES:** Gloss white Testors enamel paint iris and gloss black pupil eyes

Silver Opossum

Tied by Mike Valla

- **HOOK:** #8-10 Mustad 9674
- **THREAD:** Red 6/0 UNI-Thread
- **BODY:** Silver Lagartan oval French tinsel (medium)
- **THROAT:** Red hen hackle
- **WING:** Opossum guard hair and underfur

Southern Redbelly Dace

Tied by Mike Valla

- **HOOK:** #4-8 Mustad 3665A
- **THREAD:** Black 6/0 UNI-Thread
- **TAIL:** Green bucktail
- **RIB:** Silver tinsel
- **BODY:** Mylar tubing (medium)
- **WING:** Green bucktail over black bucktail (top) and black bucktail over white bucktail over pink bucktail (bottom)
- **HEAD:** Gloss white Testors enamel paint iris and gloss black pupil eyes

Sulphur Quill Dun

Tied by Tom Miller

- **HOOK:** Daiichi 1180 or Tiemco 100, #16-20
- **THREAD:** Yellow Danville 6/0 waxed
- **TAIL:** Medium dun or light ginger hackle barbs, splayed
- **BODY:** Stripped hackle stem dyed yellow, or yellow overdyed with a hint of orange
- **WING:** Medium dun hen hackle tips
- **HACKLE:** Medium dun or light ginger

Note: Variation of A. K. Best pattern

Blue-Winged Olive Cut Wing Thorax Dun

Tied by Tom Miller

- **HOOK:** #18-22 Daiichi 1180 or Tiemco 100
- **THREAD:** Olive Semperfli Nanosilk 18/0
- **TAIL:** Medium to dark dun hackle barbs, splayed
- **BODY:** Dark olive beaver, heavy at thorax
- **WING:** Dark dun feathers
- **HACKLE:** Medium to dark dun, or olive grizzly and dun

Note: This is a variation of Vince Marinaro's classic pattern. For #24-28 use Partridge K1A.

Vermont Rainbow Tandem

Tied by Scott Biron

- **HOOK:** #4-6 Mason Multibraid SS Wire
- **THREAD:** Of choice
- **BODY:** Embossed silver tinsel, oval white beads with 2 white round beads in between
- **THROAT:** White bucktail (nylon sometimes used by Ora Smith)
- **WING:** Light rose and dark rose bucktail; 4–6 peacock herls
- **CHEEK:** Jungle cock eyes
- **HEAD:** Black thread

Willoughby Lake Smelt Tandem

Tied by Scott Biron

- **HOOK:** Mason Multibraid SS Wire, #4-6
- **THREAD:** Of choice
- **BODY:** Flat silver tinsel double-wrapped and lacquered
- **THROAT:** Long white marabou
- **WING:** 3–4 peacock herls; sparse orchid hair, mixed with pearl Krystal flash; 3 light blue dun saddles with light gray grizzly on the outside (2 saddles per side)
- **SHOULDER:** Mallard breast with short jungle cock eyes
- **HEAD:** Black thread

Note: Pattern is based on Jim Warner's 1994 pattern.

Willoughby Rainbow Tandem

Tied by Scott Biron

- **HOOK:** #4-6 Mason Multibraid SS Wire
- **THREAD:** Of choice
- **TAIL:** Teal
- **BODY:** Embossed silver tinsel
- **THROAT:** White bucktail sparse and red hen
- **WING:** Yellow followed by lavender bucktail topped with several peacock herls over
- **SHOULDER:** Teal
- **HEAD:** Black thread with yellow eye and black pupil

Willoughby Smelt Tandem

Tied by Scott Biron

- **HOOK:** #4-6 Mason Multibraid SS Wire
- **THREAD:** Of choice
- **BODY:** Flat silver tinsel
- **WING:** Red, yellow, and lavender bucktail topped with 6 peacock herls over
- **SHOULDER:** Mallard flank
- **CHEEK:** Jungle cock eyes
- **HEAD:** Black thread with painted yellow eye and black pupil

Willoughby Willie

Tied by Scott Biron

- **HOOK:** Of choice, 6X or 7X long, #2
- **THREAD:** Of choice
- **TAIL:** Hot orange calf tail
- **BODY:** Fluorescent orange wool or floss
- **RIB:** Medium oval silver tinsel
- **UNDERWING:** Red bucktail
- **WING:** Single straight gadwall spear feather tied in flat
- **THROAT/COLLAR:** Hen grizzly tied back
- **HEAD:** Red thread with yellow or white eye and black pupil
- **SHOULDER:** Wood duck flank
- **CHEEK:** Jungle cock eyes
- **HEAD:** Black thread

Note: Based on Mike Martinek's pattern of Joe's smelt for northern Vermont fall fishing. Named for Bill Lockwood, owner of Bill and Billies Lodge on Lake Willoughby.

Index

Ace Brook, 124, 127
Adams, Bill, 13
Adams Parachute fly, 220
Adams Pool, 13–14
Aiken, George D., xv
Alan, R. Strobridge Recreational Complex, 29, *30*
Alder Brook, 17
Alder Meadow Brook, 102
Alexopolous, Chris, 41
Al's Little Perch fly, 159, *159*, 256
Amadon, Clark, 105
American eels, 56
American Museum of Fly Fishing, 5, *6*
Angler's Trail, *96*, 97
annual fisheries assessments, xviii–xix
A.P. Black Beaver flies, 51, *51*
Appalachian Trail (AT)
  Big Branch Stream and, 59–62, *61*, *62*
  City Stream and, *27*, 28
  Clarendon Gorge trailhead, 75
  Stratton Mountain Pond area, 243
Arlington, 5
Arnold Falls Dam, 199
AT. *See* Appalachian Trail
Atherton, John, "Jack," 11
Atherton, Maxine, "Max," 11
Atherton Number 1 fly, 165, *166*
Atherton Number 2 fly, 165, *166*
Atocha, Marion, 78
Atocha, Steven, 77–78
Ausable Wulff fly, 245, 256
Averill Lakes, 170, *172*, 173–75, *174*, *175*

Bailey Brook, *214*, 214–15
Bald Hill Fish Culture Station, 149
Bald Hill Pond, 149–50, *150*
Baldwin Creek, 84–85
Ball Mountain Reservoir, 246
Baltz, Tom, 14, 188
Barton River, 154
Bartonsville Covered Bridge, 249
Basin-1 Tactical Basin Plan, 2
Basin 7 Tactical Basin Plan, 109, 113
Basin 10 Tactical Basin Plan, 209, 212, 214, 216, 230
Basin 14 Tactical Basin Plan, 225
Basin 17 Tactical Basin Plan, 142
Bates, Joseph D., Jr., xv
Batten Kill, *xiv*, xv, xxi, 2–5
  Arlington to New York border reaches, *3*, 10–11
  brook trout and, 4, 9, 12
  brown trout and, 4, *7*, 9
  hatches, 14–16
  headwaters and west branch to Manchester, 8–9
  Manchester area, 5
  Manchester to Arlington reaches, *9*, 9–10
  Mettawee differences from, 40
  multiple-use recreation, *4*, 11–12
  New York state reaches, 11–14
  Walloomsac differences from, 23
The Battenkill (Merwin), 4
Batten Kill Badger fly, 14
Battenkill Flats fly, 256
Battenkill Museum and Park, 13
Bean Brook, 196
Beard Recreation Park, 120
Bear Mountain, 4
Beaver Brook, 47
Beaver Kill, xxi
Before Pool, 163, *163*, 165
Beldens Falls, 63
Beldens Hydroelectric Station, 63
Bennett, Bob, 68
Bennington Battle Monument, 22–23, *23*, 26
Bennington Fish Hatchery, 242
Bennington Wastewater Treatment Plant (WWTP), 25
Benoit, Chloe, *86*
Benoit, Ted, 151–53, 159, 197, 198
Bickford Hollow Brook, 22
Big Black Branch, 60
Big Branch Stream, *59*, 59–62, *61*, 62
  Otter Creek and, 64
Big Falls, 128, *128*
Big Fish Fly pattern, 222, *223*, 256
Biron, Scott, *xv*, *142*, *158*
Black Avenger fly, 241
Black Branch, 185
Black Brook, 17, 20, 234, 242–43, *244*
Black Creek, 4
Black Falls Brook, 136
Black Pond, 209
Black River, xviii, *209*, 209–12, *210*, *211*, 249
  tributaries, 212
Blue-Winged Olive Cut Wing Thorax Dun fly, 261
Blue Wing Olive mayflies, *xxiii*, 15, *15*, 177
boat portage markers, 242
Bolles Brook, 22, 26–29, *27*
Bolton Dam, 86
Bolton Falls, *89*, 89–92
Bomoseen State Park, 56
Bornholdt, Dave, *210*
Bornholdt, Pamela, *210*
Bornholdt Golden Squirrel fly, *212*
Bourn Pond, 2, *20*, 20–22
bowfin, xi, *xi*, 36–38, 131
Boyd, Morgan, *27*
Bradford Dam, 225
Branch Pond, 2, *19*, 20–22
Branch Pond Brook, 17

Brandon waterfall, 71–72
Brandy Brook, 77
Breadloaf Wilderness, 202
Brew's Bugger flies, *39*
Bridge of Flowers, 130
Bridgewater Park, 231
Brockways Mills Gorge, 249
brook trout, xiii, *xiii*, *xiv*, 2
  Bailey Brook and, 215
  Batten Kill and, 4, *7*, 9, 12
  Big Branch Stream and, *60*
  Black River and, 210
  Browns River and, 122
  Colton Pond and, 232
  Connecticut River and, 183
  Deerfield River and, 242
  Dog River and, 107
  Forest Lake and, 170
  Gihon River and, 121
  Great Averill Lake and, 174
  Jobs Pond and, 148–49
  Kenfield Brook and, 115
  Kent Pond and, 233
  Knapp Ponds and, *215*, 216
  Lamoille River and, 112
  Mad River and, 102, 103
  Martins Pond and, 223, 224
  Mettawee River and, 41
  Missisquoi River and, 125, *126*
  Neshobe River and, 70
  Notch Pond and, 188
  Ompompanoosuc River and, 227, *229*
  Ottaquechee River and, 230, 231
  Passumpsic River and, 196
  Paul Stream and, 190
  Pherrins River and, 143, *146*
  Pine Brook and, 146
  Roaring Branch and, *18*
  Roaring Brook and, 231
  stocking ponds with, *20*, 21
  Sugar Hill Reservoir and, 72
  Trout River and, *138*, 138–39
  Walloomsac River and, 23
  White River and, 204
Brouillard Brook, 187
Brown, Roy, 12
Brown Bullhead flies, 217
Browns Covered Bridge, 73–74, *74*
Browns River, 121–23, *122*
brown trout, *xiv*, 2
  Bailey Brook and, 215
  Batten Kill and, 4, 9
  Black River and, 210, 211
  Castleton River and, 54–55
  Connecticut River and, 183
  Deerfield River and, 242
  Dog River and, 107
  Forest Lake and, 170
  Furnace Brook and, 68
  Gihon River and, 121
  Hoosic River and, 33
  Kenfield Brook and, 115
  Lake Bomoseen and, 56
  Lamoille River and, 111
  Little River and, 95–96
  Mad River and, 102
  Missisquoi River and, 128
  Otter Creek and, 65
  Passumpsic River and, 199
  Poultney River and, 50
  Trout River and, 138
  Walloomsac River and, 23, 24
  Waterbury Reservoir and, 95
  Winooski River and, 87, 88
Buckley, Courtney, 212, 214–17
Buffum's Bridge, 13
Bugeja, Paul, 163, *163*, 165
Bull Run, 105, 106
Burgess Branch, 123, 126

Cadoret, Brian, "Lug," 63
Cady's Falls, 114
Cage Dam, 199
Camden Creek, 4, 12
Camp Plymouth State Park, 210
Canaan, 175–76, 178–82
canoeing, 11–12
Card, Annie, 29
carp, 131, *132*, *133*
Carp Bug fly, *133*
Carpenter Ant flies, 52, *55*
Carver Falls, 49, 51
Caspian Lake, 112, *112*, 156, 158
Castleton River, xv, 49, *52*, 52–55, *53*
Catamount Trail Cross Country Ski markers, 242
Catamount X-C Ski Trail, 239
Cavendish Dam, 210, 211
Cayuga Lake, 165
CCC. *See* Civil Conservation Corps
CDC Grannom Pupa fly, 14, *14*, 188, 256
Cemetery Road reach, *53*, 54
Chamberlain, Keith, 152
Champlaign Canal, 45, 46
de Champlain, Samuel, 36
Champlain Special Tandem fly, 256
Chenango River, 221
Cherry Peel Optic Bass fly, 257
Chocklett, Blane, 36
chop and drop practices, 188
City Stream, 22, 26–29, *27*
Civil Conservation Corps (CCC), 96
Civil War, 72
Clarendon Gorge, 75
The Classification of Aquatic Communities of Vermont, 52
Cleveland Brook, 207
Clyde Pond, 162
Clyde River, xvi, *140–41*, 142, *158*
  lower river landlocked salmon section, 160–65, *161*, *162*, *164*
  Special Regulations, 167
  upper, 166–67, *167*
Coaticook River, 147, *147*
Coffin Fly, 222

Cold River, 73–74, *74*
Cold Spring, 22
Cold Spring Brook, 24
Collins, Ed, 220
Collins, Laurie, 220
Colton Pond, 232, 233, *233*
Comfort, Brian, 240–41
Connecticut Lakes, 176
Connecticut River, xvi
  license regulations for, 179
  regulations on, xix
  Saxtons River and, 252
  Upper, 174, 175–83, *176*, *180*
  Wells River and, 217
  White River and, 208–9
  Williams River and, 252
Connecticut River Basin, 2, 170
Connecticut River Paddlers' Trail, 181
Connecticut River Streambank Access, 178, *178*, *179*
Cooley Covered Bridge, 68
Coolidge Range, 73
Cooper's Brook, 113
Copper Kettle fly, 159, *159*, 160, 257
Cox Brook, 108
Craine, Joe, xvi
Cross Brothers Dam, 107
Crossett Brook, 88
Cutler, Isabella, 41
Cutler Memorial Forest, 41, *41*, 42

Damm, Brett, 222
Darbee, Elsie, 68
Davis Park, 129
Deerfield River, xviii, 2, 20, 170, 234, *235*, *236*
  Black Brook tributary and Upper East Branch, 242–43, *243*, 245244
  headwaters and main branch, 241–42, *242*
  Readsboro section and West Branch, 238–41, *240*, *241*
  Searsburg bypass, 236–38
DeForge Hydroelectric Station, *89*
DeLorme: Vermont Atlas & Gazetteer, xviii
Devlin, Gene, 171, 173
Dietrich, Willy, 109, 111
Dionondahowa Falls, 4
Dish Mill Brook, 196
DNA sequencing, xvi–xvii, 8–9
Dog River, 88, 103, 105–9, *106*, *108*
Dog's Head Falls, 115–16
Dog Team Falls, 83, *84*
Dorset Hollow, 41
Dorset Mountain, 41
Drew's Bugger fly, 257
Dutchman's Hole, 12
Dwight D. Eisenhower National Fish Hatchery, 67–68, 242

Eagleville Covered Bridge, 13
East Barnet Hydro Plant, 200, *200*, 201
East Branch Trail, 242
East Creek, xviii, 63, *66*, 66–67
Eastern Fly Fishing (magazine), 14
Echo Lake, xvi, 157–60, *158*, *159*, 209, 210
Edson Tiger flies, 24, *26*
Edson Tiger-light fly, 257
Edwin Wright Orvis Memorial Park, 85
Eisenhower, Dwight D., 68
electrofishing devices, xvi
  trout population surveys with, *5*
Elk Hair Caddis flies, 229
Elmore Branch, 113–14
Emerald Lake, 62, 64
Emerald Shiner flies, *187*
Emily's Bridge, xix, 98, *99*
Enosburg Falls, 123, 130, *130*
Ephemerella dorothea dorothea, 14, 24
Ephemerella invaria, *xxiii*
Ephemerella subvaria, *xxiii*, 11, 24, 107
Equinox Hotel, 5, *6*, 9, 10
Estaz Egg flies, 79, *80*
Eurasian water milfoil, 56

Fall Creek, 165
Fathead Minnow flies, *187*, 217, 259
Federal Energy Regulatory Commission (FERC), 162, 237
Female Rusty DNA spinner, 257
FERC. *See* Federal Energy Regulatory Commission
Fife Brook Dam, 234
First Congressional Church, 5, *6*
First Connecticut Lake, 176
fishing regulations
  Connecticut River and, xix, 175, 179, 188
  updates to, xvii
  White River and, 206–7
fish ladders, 162
fish-passage facilities, 162
flash flooding, 195
Float Bridge, 57
Flower Brook, 44, 47, *48*, 49
Flower Brook Gorge, 47
Fly and the Fish (Atherton, J.), 11
The Fly Fisher and the River (Atherton, M.), 11
Fly Fishing the Hex Hatch (Wass), 149, 154, 220
Forest Lake, 170–71, 174
Fourth Connecticut Lake, 176
Fred Mold Park, 199
Freight Train fly, *117*
Friends of the Winooski River, 106
Frost, Robert, 77, 151
Fuller, Roger, 72
Furnace Brook, 22, 63, *67*, 67–69

Gallup Mills, 193, 194
Game Changer flies, 37, *38*
gar, xi
Georgi, 13
Gerardi, Leonard, 170–71
Gifford Woods State Park, 233
Gihon River, *110*, 114, *119*, 119–21, *120*
Gilchrist, Charles, "Pook," 13
Glastenbury ghost town, 26, 28

GMP. *See* Green Mountain Power
Gold Brook, xix, 98, *99*
Golden Darter fly, 19, *19*, 257
Golden Squirrel fly, *212*
Golden Stonefly nymph flies, 197, *200*, 257
Good, Shawn, 55–57, 70, 72
Gorham Covered Bridge, 68
Goshen Brook, 77
Goshen Dam, 72–73, *73*
Governer Aiken Bucktail fly, xv, *xv*, *150*, *158*, 258
Grannom caddis, 14
Grannom Pupa flies, 188, 256
Granville Notch, 102
Great Averill Lake, 170, *172*, 173–74, *174*
Great Bay Hydro Dam, 162
Great River Hydro power company, 241
Greenbanks Hollow, 202
Greendale Brook, 212
Greendale Campground, 248
Green Drake flies, 222
Greene, Joseph C., 83
Green Mountain Audubon River Trail, 95
Green Mountain Boys, 23
Green Mountain Hydroelectric, 51
Green Mountain National Forest, xiii, xviii, 5, 20, 41, 241, *245*
  Greendale Campground, 248
  Hancock Overlook Observation Site, 205
  mud season and roads in, 248
  Riverbend Observation Site, *204*, 206
  sign-in books, 61
Green Mountain Power (GMP), 72–73, 116
Green Mountains, xv, xvi
Green River, 16–17, *17*
Greensboro Bend, 112
Grendale Brook, 246, *246*, 248
Griffith Lake, 59, 61
Grocery Pool, 12
Grout Pond, 2, 20–22, *21*

habitat enhancement, 4
Hancock Overlook Observation Site, 205
Hanlon, Shane, 68
Hannah Clark Brook, 136
Hare's Ear nymph fly, *117*
Harriman Reservoir, 234, 236–38, *238*, *239*
Harriman Station Powerhouse, 240
hatches, xxi, *xxii*
Haylon, Jeff, 72
Hendrickson CDC Thorax Dun fly, 258
Hendrickson hatch, 14, 15, *15*, 16, 45
Hendrickson mayflies, *xxiii*, 11, *11*
Henry Covered Bridge, *22*, 24
Hexagenia limbata, *xxi*
  Caspian Lake and, 112, *112*
  Echo Lake and, 157, 158
  Forest Lake and, *173*
  Jobs Pond and, 149
  Little Rock Pond and, 61
  Noyes Pond and, 220, 221, *221*, 222, 223
  Shadow Lake and, 156, *157*
  Willoughby Lake and, 153, 156
Hill Farm Riverside Conservation reach, 10
Hine, Lewis, 29
Holden Federal Fish Hatchery, 68
The Holes, 137
Hoosic River, xix, 2, 22, *28*, 29–33
Horse Pond, 109, *111*, 112
Hudson River
  Batten Kill and, 4
  Hoosic and Walloomsac Rivers and, 22, 29
Hudson River Drainage Basin, 2
Huntington River, xviii, 91–95, *94*
hydroelectric facilities, xviii, *xviii*

Inland Fisheries report, 223, 252
  on Echo Lake, 157
Inland Salmonid Report (Will), 215
Iron Bridge Loop, 144, *145*
Island Lake, 142, *144*
Island Pond, 142, 144, 166, 175
Isonychia bicolor mayflies, *43*

Jamaica State Park, *245*, 246
Jay Brook, 136–38, *137*
Jenny Coolidge Brook, 212, 246, *247*, 248
Jewell Brook, 212, *213*, 214
Jewett Brook, 22
Jobs Pond, *148*, 148–49, 224
Joe's Brook, xx, *201*, 201–2
Jonah Ventures, xvi
Juggernaut 2.0 fly, *76*, 258

Kaufmann, Randall, *117*
kayaking, 3, 11–12, 76
Kearney, Samantha, *85*
Keelan, Bill, *3*
Keelan, Don, 10
Keelan reach, *3*, 10
Kenfield Brook, 114–15
Kennebago Muddler flies, 222, *223*
Kent Brook, 230
Kent Pond, 230–34, *234*
Killington Ski Resort, xi
Kingdom Bike Trail, 197
Knapp Brook Wildlife Management Area, 215
Knapp Ponds, 214, *215*, 215–17, *216*
Kratzer, Jud
  on chop and drop practices, 188
  on Clyde River, 166–67
  on Echo Lake fishery, 157
  on Moose River brook trout, 193
  on Passumpsic River brook trout, 196–98
  on pond brook trout populations, 148, 223
  Strategic Wood Additions habitat project, 187, 188, 192
  on Willoughby River rainbow trout, 156
Krieger, Margaret, 29
Kucharek, Ralph, 79–81, 93, 117

Ladago, Bret, 86, 101, 103, 107
Lake Amherst, 209
Lake Bomoseen, *xi*, 53–58, *56*, *57*
Lake Champlain, xi, *xi*, 45, 202

Lamoille River and, 109
middle section, 36–39, *37–39*
north section, 131, *132*, 133, *133*
Otter Creek confluence in, 63
Winooski River and, 86
Lake Champlain Drainage Basin, 36
Lake Champlain Restoration Program, 67–68
Lake Eden, 119, 121
Lake Francis, 176, 177
Lake George, 36, 37
Lake Memphremagog, xv, 79, *140–41*, 142, 154, 163
landlocked salmon and, 160
Lake Rescue, 209, 210
Lake Seymour, *158*, *159*
Lake Shaftsbury, 2
sea lamprey control in, 93
Lake Shaftsbury State Park, 2
Lake St. Catherine, 58–59
Lake Taker streamer, 37
lake trout, xi, 36–38, *38*
Echo Lake and, 157, 158
Lake Willoughby, xvi, 142, 149, 151–54, *152*
Lamoille River, xviii, 36, 87, 109, *109–11*, 111–17, *116*
North Branch, 117–18, *188*
Lamoille Valley Rail Trail, 113, 115
largemouth bass, *xi*, 133
Lake Bomoseen and, 55, 57
lean-to shelters, 62
Le Clair Brook, 127
The Ledges, 208
Leicester Hollow Brook, 69
Lewis Creek, 36, 78–82, *79*, *81*
Lewis Pond, 185
Lilliesville Brook, 206
Lily Pond, 58
Lime Sally stoneflies, 229, *229*, 243, 251, 258
Lincoln Covered Bridge, 232, *232*
Lion's Club Family Park, 241
Litobrancha recurvata, 223
Little Averill Lake, 170, 174–75, *175*, 185
Little Black Branch, 60
Little Black Stonefly, 51, *51*
Little Brook Trout bucktail flies, 24, *26*, 258
Little Hoosic River, 31, *31*, 33
Little Lake, 58
Little Mad Tom Brook, 4
Little Otter Creek, 36
Little River, xviii, 89
below Waterbury Reservoir, *95*, 95–97
West Branch, 97–100, *98*, *99*
Little Rock Pond, 59, 60
Lively, Chauncy K., 52
logging, 184
Long Pond, 150–51, *151*
Long Trail, 28, 59
Lucas Brook, 129
Lye Brook Wilderness, 2
Lyman Falls, 177
Lyman Falls State Park, 180–81, *181*
Mach, Gib, 47
Mach's General Store, 39, 47
Mad River, 88, *101*, 101–5, *104*, 202
Mad Tom Brook, 4, 8, *8*, 9
Major, Tim, 204
Malletts Bay, 109
Manchester, 9–10
Batten Kill near, 5
Orvis and, 41, 59
Manchester Center, 8
Manocchia, Adriano, 1, 3, *3*, *18*, 18–19
maple syrup industry, xi
Maple Syrup nymph flies, 198, *200*, 258
Marble, Tom, *95*, 97
March Brown mayflies, 177, 211
Martins Pond, 148, 223–24, *224*
Masked Avenger fly, 258–59
Masked Avenger single fly, 259
Master Angler Program, 38
mayflies, xxi, *xxi*
McClane, A. J., 19
McWilliam Covered Bridge, 255
Meade, James, 65
Meade's Falls, 65
Meadowsend Timberlands, 252, *254*
Memphremagog Basin, 142, *142*
Merwin, John, 4
Mettawee Falls, 46, *47*
Mettawee River, xv, *40*, 40–47, *41*, *43*, *44*
Lake St. Catherine and, 58
Mickey Finn Bucktail fly, 149, *149*
Middlebury Falls, 63, *63*
Middlebury Gorge, 76, 77
Middlebury River, xx, *76*, 76–78
Middle Falls, 4
Middletown Springs Park, 50
milfoil, 56
Mill Brook, 104, 122, 196, 208
Bailey Brook and, 214
Mill Brook Creek, 58
Mill Brook Culvert Fish Ladder Site, 208
Miller, Tom, 11, 15
Mill Pond, 214
Mill River, 74–75, *75*
Mills Riverside Park, 122
Milton Independent (newspaper), 116
Mineral Spring, 127
Missisquoi Bay, 36, 38, 131, *132*, 133, *133*
Missisquoi River, xviii, 36, 123, *123*
East Branch, *124*, 124–26, *125*, *126*
main stem, 126–31, *127*, *128*
Trout River and, 134, *134*
upper section, *135*
water temperatures, 126–27
Wild and Scenic River designation, 129
Missisquoi Valley Rail Trail, 129
Montgomery Center, 134, 136
Montgomery Covered Bridge, 117–18, *188*
Montreal White Belly tandem fly, 259
Moore Park, 122
Moosalamoo National Recreation Area, 77
Moose River, xviii, 192–95, *193*, *194*, 199

Moretown No. 8 Hydro Dam, 105
Morgan Covered Bridge, 118
Most Terrifying Places in America (television series), 98
mountain biking, xi, 97, 98
Mountain Brook, 47
Mount Equinox, 5, 16
Mount Holly, *245*
Mount Mansfield State Forest, 98
Mr. Bow Jiggles flies, 37, *39*, 259
Murphy Dam, 176, 180
MV Cossayuna Fathead Minnow fly, 259
MV Emerald Shiner fly, 259
MV Mettawee Rainbow fly, 42, *42*

National Recreation Trail Loop, 77
Native Fish Coalition (NFC), xi, 163, *163*
The Nature Conservancy, 116
Neal, Tim, 220
Neshobe River, 63, *69*, 69–72
Newark Pond Natural Area, 150
New Hampshire
    Connecticut River regulations and, xix, 175, 179
    Connecticut River Streambank access in, *178*, 178–79
    Upper Connecticut River stretches in, 176–77, 179–80
New Haven River, 63, 76, *82*, 82–85, *84*
Newhouse, Jeff, *162*
Newport, *140–41*, 163
New York
    Batten Kill reaches in, 11–14
    Hoosic River stretches in, 31, *31*, *32*, 33
    Mettawee section in, 45–47
    Walloomsac reaches in, 25–26
NFC. *See* Native Fish Coalition
NFCT. *See* Northern Forest Canoe Trail
Nine Island, 195, 200
North Breton Brook, 53
North Brook, 50
Northeast Kingdom, xv–xvi, xx, 170, 197
    floods in, 195
Northern Forest Canoe Trail (NFCT), 129, 181
Northern Lake Champlain, 36
North Ferrisburgh Falls, *79*
North Springfield Reservoir, 209
North Street Dewey Recreational Fields, 54
Norton Pond, 143, 147, *147*
Notch Pond, 188–90, *189*
Noyes Pond, 218, *220*, 220–23, *222*
Nulhegan Basin, *185*
Nulhegan River, xvi, xx, xxi, *184*, 184–88, *186*, *187*
    Connecticut River and, 182, *182*
Number 11 Dam, 160

Oatman, Lew, 12, 13, 19
Okemo State Forest, 212
Ompompanoosuc River, xx, 170
    East Branch, *226*, 226–27, *227*
    West Branch, *228*, 228–29, *229*
open-season fishing dates, xvii
Orphan fly, *166*
Orvis, 5, *7*, 41, 59, 68
Ottauquechee River, 170, *230*, 230–32
Otter Creek, xviii, *34–35*, 36, 37, 62–66, *63–65*
    Big Branch Stream and, 59, 60
    Middlebury River and, 76–77
    New Haven River and, 83–84
    tributaries, 66–72
Otter Creek Wildlife Management Areas, 64
Owl Kill, 33
Oxbow Riverfront Park, 114

panfish, 57
Passumpsic River, xviii, 193, *194*, 195
    East Branch, *196*, 196–97
    main stem, *199*, 199–200
    West Branch, 197–98, *198*
Passumpsic Valley Land Trust, 202
Patch Dam, 66, 67
Patriot Hydroelectric Station, 180, *180*
Patterson Brook, 204, *205*
Paul Stream, xvi, 184, 188, 190, *190*, *191*, 192
Peiser Pond, 202
Penns Creek, xxi
Pensioner Pond, 167
Peterson Dam, xviii, 109, *116*, 116–17
PFR. *See* public fishing rights
Pheasant Tail nymph fly, *117*
Pheasant Wing Hexagenia fly, *223*, 260
Pherrins River, xvi, *143*, 143–47, *145*, 166
Phillips, Colin, 102, 104, 105
pike, 63, 66
Pine Brook, 146
Pitt, Al, 152, 159, 160, 197–98
Plymouth Lakes, 209, 210
Polar Sculpin fly, 260
Pond Hill Brook, 53, 54
Pook's Bridge, 13
Potash Mountain, 185
Pottersville Dam, 113
Poultney River, 49–51, *50*
    Castleton River and, 53
Power House Covered Bridge, 119–20, *120*
Pownal Tannery Hydroelectric Power Plant, *28*
The Practical Fly Fisherman (McClane), 19
Preston Brook, 92
Price, Drew, *xi*, 36–38, *38–39*, 39, 131, *132–33*, 133
Prindle, Al, 12
Proctor Falls, 66
Proulx, Nick, 177
public fishing rights (PFR), 31
Purple Matuka streamer, *166*

Quayle Conservation Area, 103, *104*
Quimby, Charles, 173
Quimby, Hortense, 173
Quimby Country, 170–75

rainbow smelt, 36
rainbow trout, *xi*, xv, *xvi*, xvii

Bald Hill Pond and, 150
Connecticut River and, 183
Deerfield River and, 238, *238*
Dog River and, 107, 109
East Creek and, 66
Echo Lake and, 157, 158
Enosburg Falls and, 130
Flower Brook and, 49
Furnace Brook and, 68
Hoosic River and, *32*, 33
Kenfield Brook and, 115
Kent Pond and, 233
Lamoille River and, 111, 113
Little River and, 95–96
Mad River and, 103
Mettawee River and, 42
Nulhegan River and, 188
Ottaquechee River and, 230, 231
Passumpsic River and, 199
Saxtons River and, 255
strain evaluation, xix
trolling streamers and, 152, 153
Waits River and, 225
Walloomsac River and, 23, 24
Waterbury Reservoir and, 95
Willoughby River and, *155*, 156
Winooski River and, 88, *91*, 92
Ramsay, Henry, 11
Ranch Brook, *xiii*, *100*, 100–101
Recurve Hatch, 223
Red Brook, 217
Redfin pickerel, 57–58
Red Quill DOA Cripple Emerger fly, 260
Red Quills, *11*
Rescue Lake, 209
Richford Conservation and Access Area, 129
Richmond, John, 72
Ricker, Charlie, 96
Ricker Pond, 217, 218, *218*
Ridley Brook, 90
Riverbend Observation Site, *204*, 206
River Walk Park, 134
Roaring Branch (Batten Kill), 1, 17–19, *18*
Roaring Branch (Walloomsac), 22, 24–29
Roaring Brook, 231
Rockwell, Norman, 11
Rogers, John, 29
Round Mountain, 185
Royal Trude fly, 111

safety precautions, xviii
salmon, xv
Connecticut River and, 177, 196
hatchery for, 67–68
landlocked, 92–93, 160–65, *162*, *165*, 177
trolling streamers and, 152
Winooski River and, *92*, 92–93
Salmon Hole, *92*, 93
Salmonid Inventory and Management reports July 1, 20222-June 30, 2023, 148
Sandbar Smelt tandem fly, 260
Sargasso Sea, 56
Saxtons River, 249, 252, *253–55*, 255
Schmidtmann-Valla smelt fly, 245, 260
sea lampreys, 93
Searsburg Reservoir, 241, *242*
dam spillway, 236–38, *237*
Second Connecticut Lake, 176
"A Servant to Servants" (Frost), 151
Sewer Hole, 14
Seyon Lodge, *220*, 221
Seyon Pond, 153, 154, *158*, *220*, 220–23, *222*
Seyon Ranch, 220, *221*
Shadow Lake, 156, *157*
Shaggin' Draggin' fly, *133*
Shelburne Falls, 234
Sherman Reservoir, 234, 240
Shumlin, Peter, xv
Shushan Postmaster bucktail fly, 12, 33, 42, *42*
sign-in books, 61
Silver Opossum fly, *212*, 260
Silvio O. Conte National Wildlife Refuge, xx, 184, *185*, 189
Simanon, Lee, 115
Simard, Lee, 118, 123, 125–27, 136
skiing, xi
Skylight Pond, 202
Slate Valley Trail System, 51
SMA. *See* Streambank Management Area
smallmouth bass, *xi*
Lake Bomoseen and, 55, 57
Otter Creek and, 66
Smith-Root eDNA sampling unit, xvi, xvii, 242
Smugglers' Notch, 98
Snider Brook, 126, 127
Somerset Reservoir, 234, 242, 243
South Bay, 49
Southern Lake Champlain, 36
Southern Redbelly Dace fly, *189*, 260
South Stream, 22
Spaulding Brook, 187
Special Regulations Trophy Trout sections
Black River, 210, *210*
Deerfield River, 241–42
Dog River, 105, 108
East Creek, 66
Enosburg Falls, 130, *130*
Martins Pond, 223
Missisquoi River, 123, 130, *130*
Moose River, 193
Otter Creek, 64
Passumpsic River, 199
Upper Connecticut River, 175, 176, 180, 182, 188
Walloomsac River, 23–25
White River, 206–7
Winooski River, 88–89
spinner falls, *xxiii*
Spirit of Pittsford Mills fly, 68, *69*
Sprague, Myron, *166*
Spring Hole, 12–13
Spruce Mountain, 220
S&R Hexagenia fly, *223*
S&R Hex Spinner, 220

Stamford Stream, 22
Stannard Brook, 112
Stark, John, 23, 26
Stark, Molly, 23
State Fishing Regulations, xi
  Connecticut River and, xix, 175, 179, 188
  updates to, xvii
  White River and, 206–7
steelhead, xi
  Echo Lake and, 157
  Lewis Creek and, 78–81
  Memphremagog Basin and, 143
  Willoughby River and, 154, *155*, 156
  Winooski River and, 93
Steen, Stan, 211, *211*
Stetson Brook, 102, 103
St. Lawrence River, 56
stocking
  Browns River, 122
  Castleton River, 54–55
  Colton Pond, 232
  East Creek, 66
  Enosburg Falls, 130
  Forest Lake, 170–71
  Harriman Reservoir, *238*
  Knapp Ponds, *215*
  Lake Bomoseen, 56
  Lamoille River, 111
  Lewis Creek, 79
  Mad River, 103
  Martins Pond, 223
  Neshobe River, 70
  Nulhegan River, 188
  Otter Creek, 65
  Passumpsic River and, 199
  ponds, *20*, 21
  put-and-take, 134
  steelhead, 79, 157
  Sugar Hill Reservoir and, 72
  Trout River, 134
  Waits River, 225
  Walloomsac River, 23
  Waterbury Reservoir, 95–96
  Wells River, 217
  Williams River, 249
  Winooski River, 88, *91*
Stowe, 101
Stowe Land Trust, 100
Stowe Mountain Resort, xi
Stowe Recreation Path, 97
Strategic Wood Additions habitat project, 187, 188, 192
Stratton Mountain Pond trailhead, 243
Stratton Mountain Ski Resort, 245
Stratton Pond, 243
Streambank Management Area (SMA), 81–82
Streamer Fly Tying and Fishing (Bates), xv
streamers, trolling, 152–53, *153*
Streeter, Rob, 37
Strolis, Rich, 240–41
Sucker Brook, 73
Sugar Hill Reservoir, 72–73, *73*
Sugar Hollow Brook, 68
sugarhouses, xi
sugar maples, xi
Sulphur Quill Dun flies, *26*, 261
Swanton Dam, 130–31, *131*
Sztorc, Scott, 3, 16

Tabor Branch, 225
Taconic Range, 5
Ten Bends, 115
That Beautiful Vale above the Falls (Yaeger), 96
Third Connecticut Lake, 176
Third Hole, 137
Thundering Falls, 231
tools, xviii–xix
topological maps, xviii
Townshend Lake, 246
Trainor, J. T., 13
Trico hatch, 10, 14
Tricorythodes spinners, *9*, 10
trolling streamers, 152–53, *153*
Trophy Trout Waters, xviii
Tropical Storm Irene, xx
  Dog River and, *106*, 107
  Flower Brook and, 49
  Mad River and, 102
  Mad Tom Brook and, 8
  Mill River and, 74
  Roaring Branch (Batten Kill) and, 18
  Roaring Branch (Walloomsac) and, 26
  Williams River and, 248
Trout River, 129, 134, *134*, *135*
  South Branch, *138*, 138–39, *139*
  tributaries, 136–38
Trout Unlimited, xi, 3, 188, 208
Truthville reach, 45–46, *46*
tubing, 3, *4*, 11–12
Tutalo, Ken, 30
Tweed River, 206
2017 Basin 8–Winooski River Watershed Water Quality and Aquatic Habitat Assessment Report, 87, 99
Tying the Founding Flies (Valla), 19

Umberger, Rose Mae, 122
Union Street Fork, 9
Union Village Dam, 227
Union Village Dam Multiuse Trail Network, 227, 228, *228*
Unknown Pond, 185
Upper Connecticut River, 174, 175–83, *176*, *180*
  Bloomfield, 182–83
  Canaan to Bloomfield, 179–80
  Canaan to Connecticut River Streambank Access, 178–79
  Colebrook access, 181
  Lyman Falls State Park, 180–81, *181*
  New Hampshire stretches of, 176–77
Upper Missisquoi and Trout Rivers Wild and Scenic Committee, 129
Upper Winooski River Fisheries Assessment (Ladago), 86

US Army Corps of Engineers, 96, 227, 246
USDA Forest Service, 241
US Fish and Wildlife Service, 188

Vergennes Falls, *34–35*
Vermont
  DNA analysis of streams in, *xvi*, xvi–xvii
  hatches in, xxi, *xxii*
  Trophy Trout Waters, xviii, 23–25, 64
Vermont Fishing Regulations, xi, xvii
  Connecticut River and, 179, 188
  White River and, 206–7
Vermont Natural History Report, 52
Vermont Natural Resources Council, 116
Vermont Rainbow tandem fly, 261
Vermont River Conservancy, 120
VF&WD, xi
  annual fisheries assessments, xviii–xix
  Bailey Brook and, 214
  Black River and, 210, 212
  Clyde River and, 160
  Colton Pond and Kent Pond and, 232
  Deerfield River and, *242*
  Dog River and, 107, 108
  Echo Lake and, 157
  fishing access areas, xix
  Green River and, 17
  Jewell Brook and, 212
  Jobs Pond and, 148
  Knapp Ponds and, 215, 216
  Lamoille River and, 115
  Martins Pond and, 223
  Master Angler Program and, 38
  Neshobe River and, 70
  Passumpsic River and, 199
  Ranch Brook and, 101
  trout population surveys, *5*
  Waits River and, 225
  Williams River and, 249
  Willoughby River and, 156
  Winooski River and, 87, 88
Victory Basin Wildlife Management Area, 193

Wabanaki Conservation Area, 103
Wade Brook, 134, 136–38
Waits River, 225, *225*
Wallingford Recreational Park, 65
Wallomsac River, xviii, 2, *22*, 22–26, *25*
Walloomsac River, xv
Warm Brook, 17
warmwater bass, 57
Warner, Jim, *158*
Warren Falls, 102, 103
Wass, Leighton, 149, 153, 154, 157, 220–22, *222*
Waterbury Dam, 96
Waterbury Reservoir, 95–97, *96*, 97
Waterman Brook, 116
Webster, Dwight A., xxi
Wells River, *216*, *217*, 217–18, *218*
West Branch, 4, 8–9
  of Little River, 97–100, *98*, *99*
West Charleston Pond, 167
Western Coachman flies, 84
West Mountain Wildlife Management Area, 188, *189*
West River, *245*, 245–46
Weyerhaeuser Corporation, 188
Wheeler, Dean, *176*, 183
Wheeler, Scott, 160
When Salmon Was King (Wheeler), 160
Whisker Biscuit fly, *133*
White Creek, 4
White River, *xi*, *168–69*, 170
  Bethel to South Royalton, 207–8
  headwater reaches to Rochester, 202, 204–6, *205*
  middle section, *203*
  Rochester to Stockbridge, 206
  South Royalton to White River Junction, 208–9
  Special Regulations section, 206–7
  Stockbridge to Bethel, 206–7, *207*
White River National Hatchery, 207
White Wooly Bugger flies, 79, *80*
wild boar, xvii
wild brook trout, xiii, *xiii*, *xiv*
Wildlife Management Areas (WMAs)
  Knapp Brook, 215
  Otter Creek, 64
  Victory Basin, 193
  West Mountain, 188, *189*
Will, Lael, 210, 215, 252
Williams River, 248–49, *249–51*, 251–52
Williams River State Forest, 251
Willougby Willie fly, 262
Willoughby Falls, *155*
Willoughby Lake Smelt tandem fly, 261
Willoughby Rainbow tandem fly, 261
Willoughby River, 79, 154, *154*, *155*, 156
Willoughby Smelt tandem fly, *142*, 261
Willoughby Willie streamer, *153*
Windham Foundation, 255, *255*
Winnie Park, 252, *253*
Winooski One Dam, *92*, 93
Winooski River, xviii, 36, *85*, 85–93, *89–92*, 99
  Dog River and, 106, 108
  Little River and, 96–97
  Mad River and, 102, 105, 202
  Ranch Brook and, 100
Wixom, Scott, 21, 241
WMAs. *See* Wildlife Management Areas
Wooly Bugger flies, 78, 79, *80*, 116, 207, 225
Worrall Covered Bridge, 249
Wright, Parker, 85, 86, 162, 165
Wulff, Lee, 13
WWTP. *See* Bennington Wastewater Treatment Plant

Yaeger, W. Patrick, 96
Yellow Bogs, 185
Yellow Branch, 185
Yellow Sally stonefly, 128
Yellow Stonefly, *xxiii*

About the Author

Mike Valla was inducted into the Fly-Fishing Hall of Fame in 2024. He is the author of *Fly Fishing Guide to New York State, The Founding Flies, Tying Catskill-Style Dry Flies, Tying and Fishing Bucktails and other Hair Wings, Tying the Founding Flies, The Classic Streamer Fly Box,* and *Favorite Flies for the Catskills.* He is the recipient of the Poul Jorgensen Golden Hook Award and is often invited to speak about fly fishing at local, national, and international fly-fishing events and gatherings. He is a frequent contributor to *American Fly Fishing* magazine and has published articles in *Fly Fisherman, Fly Tyer,* and *American Fly Fisher.* He served on the board of directors at the Catskill Fly Fishing Center and Museum. Valla has been fly-fishing for well over 50 years. He lives in the town of Cambridge, New York, just a few miles from Vermont.